RAND

Patterns of Enlisted Compensation

M. Rebecca Kilburn, Rachel Louie,
Dana P. Goldman

Prepared for the
Office of the Secretary of Defense

National Defense Research Institute

The research described in this report was sponsored by the Office of the Secretary of Defense (OSD), under RAND's National Defense Research Institute, a federally funded research and development center supported by the OSD, the Joint Staff, the unified commands, and the defense agencies, Contract No. DASW01-95-C-0059/TO01.

Library of Congress Cataloging-in-Publication Data

Kilburn, M. Rebecca.
    Patterns of Enlisted Compensation / M. Rebecca Kilburn, Rachel Louie, Dana P. Goldman.
      p. cm.
    "Prepared for the Office of the Secretary of Defense."
    "National Defense Research Institute."
    "MR-807-OSD."
    Includes bibliographical references.
    ISBN 0-8330-2479-5 (alk. paper)
    1. United States—Armed Forces—Pay, allowances, etc. I. Louie, Rachel, 1960–   .
II. Goldman, Dana P. (Dana Paul), 1966–   . III. National Defense Research Institute (U.S.) IV. United States. Dept. of Defense. Office of the Secretary of Defense. V. Title.

UC74.K55 2001
331.2' 81355' 00973—dc21

                                    96-54487
                                        CIP

RAND is a nonprofit institution that helps improve policy and decisionmaking through research and analysis. RAND® is a registered trademark. RAND's publications do not necessarily reflect the opinions or policies of its research sponsors.

Published 2001 by RAND
1700 Main Street, P.O. Box 2138, Santa Monica, CA 90407-2138
1200 South Hayes Street, Arlington, VA 22202-5050
201 North Craig Street, Suite 102, Pittsburgh, PA 15213-1516
RAND URL: http://www.rand.org/
To order RAND documents or to obtain additional information, contact Distribution Services: Telephone: (310) 451-7002; Fax: (310) 451-6915; Internet: order@rand.org

Preface

Military compensation is complex, consisting of over 70 different pays and entitlements, some of which are cash payments and some of which are in-kind transfers. This complexity can produce very different levels of compensation, both across and within services, for people who appear similar by most measures. Given the power that compensation can have as a management tool, it is important for DoD policymakers to understand the components of military compensation and where and how they vary.

This report is one product of a project to evaluate issues in enlisted compensation. The project has three primary outputs. The first is a data set that can serve as a flexible tool for answering different types of questions regarding enlisted compensation. The second is a report that uses the data tool to describe the sources of variation in military compensation, and the third is a report describing ways to evaluate the economic well-being of enlisted personnel in terms of civilian standards, also using the new data set. This report, which represents the second output, will interest those involved in military compensation or military personnel management.

The research for this report was conducted for the Under Secretary of Defense (Personnel and Readiness) in the Forces and Resources Policy Center of RAND's National Defense Research Institute (NDRI). NDRI is a federally funded research and development center sponsored by the Office of the Secretary of Defense, the Joint Staff, the unified commands, and the defense agencies.

Contents

Figures

Tables

Summary

Background and Purpose

Few would think that compensation varies much among different members of the military who have the same years of service. After all, basic pay is determined by a pay table common to all services, varying only by rank and years of service. But military pay is complex, consisting of over 70 different pays and entitlements, some of which are cash payments and some of which are in-kind transfers. This complexity can produce very different levels of pay, both across and within services, for people who, by most measures, appear similar.

Given the power that compensation can have as a management tool, it is important for DoD policymakers to understand the components of military pay and where and how they vary. To date, no empirical studies have been conducted to measure the difference in total compensation by individual characteristics. This study begins to fill this gap by providing a baseline description of how pay varies for enlisted members by such individual characteristics as gender, race, occupation, family composition, and Armed Forces Qualification Test (AFQT) score.[1] We did not model the sources of the relationships between compensation and individual characteristics. Instead, we documented the patterns of enlisted compensation. Nor did we perform a behavioral analysis of these patterns—that is, we do not explain why the patterns we observed exist. Rather, our documentation of the patterns is intended to be a baseline description that can serve to identify areas in which future behavioral analysis would be especially fruitful.

Valuing Compensation

To determine how pay varies, we first looked at the components of compensation. Although a service member can be compensated from among a wide range of pay and entitlement categories, we have grouped those components into nine different categories. These appear in Table S.1, which also indicates whether the compensation is cash, an in-kind transfer (e.g., housing), or both.

---

[1]A data set was created especially to carry out this and subsequent phases of the study. It contains pay and demographic characteristics for samples of every other enlisted cohort that entered between 1978 and 1990.

Table S.1

Categories of Military Compensation

Category	Nature of Compensation
Basic Pay	Cash
Special Pays	Cash
Enlistment/Reenlistment Bonus	Cash
Basic Allowance for Quarters (BAQ)	Both
Basic Allowance for Subsistence (BAS)	Cash
Housing	In-kind
Tax Advantage	In-kind
Medical Benefits	In-kind
Retirement Benefit	In-kind

We then had to determine the value of the various components. This task posed no problem for cash payment but was more difficult for in-kind transfers. For housing, we used the median value of the housing costs for service members living off base, as reported in DoD's annual *Census of Uniformed Service Members*.[2] We calculated the tax advantage by computing the amount of additional pre-tax income a person would have to earn to yield the same after-tax income. Medical benefits were valued by determining the average cost for a health insurance premium that would provide benefits similar to those in the service.

Evaluating retirement benefits was perhaps the most challenging task, because service members generally must serve for 20 years before qualifying for such benefits. So the problem was in determining what value a person gets before 20 years from a benefit that can be claimed only after 20 years. To address this issue, we took an approach that estimates the change in each year of service to the net present discounted value of the retirement benefit.

Four Kinds of Compensation

Each service member's pay is composed of components from some part of the nine categories. Not all members receive all components, and the values of the components differ across service members, resulting in substantial variation in compensation. To capture this variation, we used four different compensation measures: Basic Pay, Regular Military Compensation (RMC), Cash

---

[2]The name of this survey is now the *Variable Housing Allowance Housing Survey*. It is maintained by the Defense Manpower Data Center, Monterey, California.

Compensation (Including and Without In-Kind Housing), and Total Compensation.

We examined Basic Pay because it is the largest component of pay and is frequently regarded as the least flexible. Because RMC is such a widely used measure of service members' take-home pay, we also looked at it. We also constructed a Cash Compensation measure that is analogous to a civilian paycheck. Our final measure, Total Compensation, is the most comprehensive. It measures both cash and in-kind transfers to service members and best represents the entire benefit a service member receives.

Results

We first document how enlisted personnel are paid—that is, we indicate which components of pay make the largest contributions to amounts, or *levels*, of pay and which components explain differences in pay across individuals, or *variances*. Then we measure the patterns of enlisted compensation across individual enlisted-personnel characteristics, such as years of service, race, gender, number of dependents, occupation, AFQT score, and service.

Who Gets What and How Much of the Compensation It Accounts For

Everyone, of course, draws Basic Pay across all years of service. The amount increases consistently, but the rate of increase slows. This gradual decline in the rate of increase holds true for all other components except Basic Allowance for Subsistence (BAS), which remains relatively constant. Between 56 and 61 percent of the enlisted population receives some type of Special Pays, again across all years of service. The fraction of enlisted personnel receiving In-Kind Housing declines as years of service increase. Not surprisingly, as the number of personnel living in base housing drops, BAQ benefits rise.

Basic Pay makes the biggest contribution to the level of pay. It accounts for more than two-thirds of Cash Compensation. The next largest contributors are BAQ (5 to 11 percent) and In-Kind Housing benefits (9 to 20 percent). All other contributors account for less than 8 percent each. Turning to Total Compensation, Basic Pay still accounts for a large fraction of the compensation, making up 45 percent in the early years and declining to 30 percent by year 14. But the Retirement Benefit assumes an increasingly large role as time passes. In year 2, it accounts for over 21 percent of the Total Compensation, and over 42 percent by year 14. Of the other in-kind transfers, the Medical Benefits account for between 6 and 10 percent, and the Tax Advantage, between 1 and 2 percent.

Sources of Variation, by Component

Although Basic Pay makes a major contribution to the level of pay, it accounts for very little of the variation, or differences, in pay. Table S.2 shows which components make the largest contribution to level of pay and variation in pay for Cash and Total Compensations.

Pay Differences Across Individual Characteristics

How does compensation differ by individual characteristics? We estimated the relationship between pay and each individual characteristic, holding other characteristics constant. These characteristics do not determine pay; rather, they might be related to pay for a number of reasons, including choices that individuals make (such as whether to marry or which occupation to enter), the design of pay policies (such as larger houses for individuals with dependents), productivity differences associated with various characteristics (such as AFQT score), or other sources of correlation between pay and characteristics. We did not model the sources of the relationship between pay and characteristics; instead, we developed a baseline case of how compensation varies by individual characteristics.

Our results show that, except for years of service, Basic Pay varies little across individual characteristics. In other words, individuals with the same years of service receive just about the same Basic Pay.

Cash Compensation and Total Compensation vary more with individual characteristics, but not by a large amount. The largest differences in

Table S.2

Contributions to Level and Variation of Compensation

Type of Compensation Measure	Biggest Contributor to:	
	Level of Compensation	Variation in Compensation
Cash Compensation	Basic Pay	Enlistment/ Reenlistment Bonus
Total Compensation	Basic Pay Retirement	Retirement Enlistment/Reenlistment Bonus Housing

compensation across characteristics were for individuals with different years of service. For example, the range that an additional year of service adds is between $656 per month to Total Compensation in the Navy and $747 extra per month in the Army. The next-largest contributor to compensation is the number of dependents: Individuals with dependents receive up to 16 percent more Total Compensation than their single counterparts. While other characteristics had statistically significant relationships with compensation, their magnitude was small, generally accounting for differences in pay of less than 5 percent.

We observed a significant relationship between individual characteristics and compensation. On the whole, however, after accounting for all characteristics, we observed that the overall variation in enlisted compensation was relatively small—that is, most personnel were receiving similar average levels of compensation.

Conclusions

The average level of enlisted Cash Compensation is most influenced by Basic Pay. However, if one service member takes home more pay than another, the source is likely to be an Enlistment or Reenlistment Bonus. For Total Compensation, the bulk of the differences—other than the bonuses—is explained by noncash components, particularly the Retirement Benefit and Housing. While Cash Compensation and Total Compensation vary with individual characteristics, most enlisted personnel receive similar levels of compensation. The two characteristics associated with the biggest observed differences in compensation are years of service and having dependents. Compensation is significantly different statistically across other individual characteristics, but the magnitudes of the differences are relatively small.

Acknowledgments

Numerous individuals provided us with support on this project. We extend thanks to Mike Dove, Matt Boehmer, Debbie Davis, Terry McMillan, and Les Willis of the Defense Manpower Data Center for providing our data. Jerry Sollinger assisted us with the development of this document. We appreciate the secretarial assistance of Kevin Mason, DeeBye Meyers, and Nora Wolverton. Members of the RAND Forces and Resources Policy Center workshop—especially Beth Asch, Glenn Gotz, and Susan Hosek—provided us with useful comments. Comments from reviewers Richard Buddin and Jacob Klerman significantly improved the report. The report also benefited from the suggestions of project monitor Saul Pleeter.

1. Introduction

Background and Purpose

Compensation is recognized as a powerful tool for managing the human resources of the Department of Defense (DoD). For instance, it is well established that compensation helps manage retention, recruiting, and separation, and helps direct manpower into desired occupations and locations (Hosek and Peterson, 1990; Hogan and Black, 1991; Asch, 1993). For compensation to be most effective as a human resources management tool, it is critical to allocate it in ways that are consistent with the DoD's objectives for managing its manpower. To evaluate the alignment between compensation policy and manpower objectives, DoD compensation policymakers need to understand how compensation varies by individual characteristics.

We know of no empirical studies that have measured differentiation in compensation by service-member characteristics, even though understanding such variation is essential to achieving DoD's objectives. This report begins to fill that gap by providing a baseline description of the distribution of compensation by individual characteristics of enlisted personnel, such as gender, race, occupation, family composition, and Armed Forces Qualification Test (AFQT) score. Such information helps DoD managers to better understand current enlisted compensation patterns and should enhance their ability to assess how well the present system meets DoD manpower objectives.

First, we documented how enlisted personnel are paid by indicating which components of compensation, such as Basic Pay, Medical Benefits, and In-Kind Base Housing, make the largest contributions to members' compensation. These results reflect the degree to which service members are remunerated in various forms: in-kind benefits or cash payments, deferred benefits or contemporaneous benefits, or components that are the same for all members instead of varying by member characteristics.

Second, we measured the patterns of enlisted compensation across individual characteristics, using regression analysis. The results indicate how compensation varies by years of service, race, gender, military occupation, number of dependents, AFQT score, and other individual characteristics. These characteristics do not determine compensation; rather, they are related to compensation as a result of individual decisions such as whether to marry or enter an occupation, pay schedule differences such as bigger on-base houses for individuals with dependents, productivity differences associated with characteristics, or other sources of correlation. We did not model the sources of the relationships between compensation and individual characteristics. Instead, we documented the patterns of enlisted compensation. Nor did we perform a behavioral analysis of these patterns—that is, we do not explain why the patterns we observed exist. Rather, our documentation of the patterns is intended to be a baseline description that can serve to identify areas in which future behavioral analysis would be especially fruitful.

The value of many of the compensation components is easily determined by looking in published DoD tables, such as the Basic Pay Tables and Variable Housing Allowance Tables (Smith and Gordon, 1994). This option raises the question of why we need to collect data when it is possible to look up the levels of compensation that service members with different grades and years of service or other characteristics would obtain. However, while the compensation tables tell us the pays for which service members with some characteristics are eligible, they do not allow us to measure the distribution of components of compensation by service-member characteristics.

For instance, if we knew certain information about an individual, we could look through DoD manuals and estimate that individual's compensation. But such tables do not indicate the number of individuals receiving that level of compensation or receiving each of the components, and how the components vary by those characteristics. In addition, while the compensation tables indicate how pay varies by some characteristics—such as pay grade, occupation, and years of service—they do not describe how pay varies by other characteristics— such as gender or race—while holding other characteristics constant.

This study shows which components make up compensation and how compensation varies by characteristics. It does not indicate whether there is sufficient variation to accomplish manpower management objectives, nor does it shed light on what those objectives should be. For instance, we did not determine whether earnings ought to vary by race and gender; rather, we determined whether they vary and by how much. In this sense, the analysis documents the patterns of compensation rather than performing the analysis that would inform compensation design. Theoretical considerations and DoD's human resource management goals should inform policymakers about how compensation should be structured. This study will provide policymakers with a baseline so that they understand how the current system deviates from their ideal system.

Data and Methods

To conduct the analysis, we used a longitudinal data set created by the Defense Manpower Data Center (DMDC) especially for this project. For enlisted personnel, the data report demographic characteristics and pay information for fiscal years 1985–1993 for every other cohort (to reduce sample size) that enlisted between 1978 and 1990. We examined the contribution of different compensation components, and we used regression analysis to identify how compensation varies by individual characteristics.

How the Document Is Organized

This report is organized into six sections. Section 2 describes the components of military pay that we studied and explains what we did not include in our measures of compensation. Section 3 defines the measures of compensation we examined and shows how we valued the different components of military compensation. Section 4 describes the methods we used to measure the contribution of the components to compensation and how compensation varies with individual characteristics. Section 5 presents the results of our analysis of the variation in compensation. Section 6 summarizes the main findings. We also include six appendices that present defining information.

2. The Components of Military Compensation

Military compensation is a bundle of separate components. Not everyone receives each component, and individuals earn different amounts of each component; so the composition of members' compensation packages varies substantially. Variation in compensation across members thus results from members' receiving pay packages that are different sets of components of varying amounts.

In this section, we describe the components that combine to form military-compensation packages. More than 70 separate pays, allowances, or benefits constitute military compensation. For simplicity, we classify these components into the following nine categories:

- Basic Pay

- Special Pays

- Enlistment/Reenlistment Bonuses

- Basic Allowance for Quarters (BAQ)

- Basic Allowance for Subsistence (BAS)

- Housing

- Medical Benefits

- Tax Advantage

- Retirement Benefit

We now proceed to describe these categories in detail, after which we discuss forms of compensation we excluded and provide the rationale for their exclusion.[1]

---

[1]For a more complete description of each of the components, see the excellent discussion in *Military Compensation Background Papers: Compensation Elements and Related Manpower Cost Items: Their Purposes and Legislative Backgrounds* (U.S. Department of Defense, 1991).

Components Included in the Analysis

Basic Pay

Basic Pay constitutes the largest component of an enlistee's monthly cash compensation. Service members receive Basic Pay in accordance with their rank and years of service (YOS), as indicated in the annual Basic Pay tables. Table 2.1 reproduces part of the Basic Pay schedule reported in *The Uniformed Services Almanac* (Smith and Gordon, 1994).

Basic Pay does not vary across individuals having the same rank and YOS, but it varies considerably across YOS and rank. The avenues to Basic Pay increases are promotion to a higher rank and longevity in the military.

Special Pays

Every enlistee's monthly paycheck includes Basic Pay; only selected service members receive Special Pays. Originally, Special and Incentive Pays were designed to compensate service members for exposure to unusually hazardous working conditions. Today, Special Pays include these types of pays—such as hazardous-duty pay and hostile-fire pay—as well as pays oriented toward inducing personnel to enter and stay in career fields that would otherwise experience shortages (U.S. Department of Defense, 1991). Many Special and Incentive Pays, such as Flight Deck Duty Pay, have elements of both hazardous-duty and critical–Military Operational Specialty (MOS) justifications. These pays

Table 2.1

Armed Forces Enlisted Basic-Pay Rates, Effective January 1, 1994

Pay Grade	Enlisted Basic-Pay Rates per Month (dollars)								
	Years of Service								
	Under 2	Over 2	Over 3	Over 4	Over 6	Over 8	Over 10	Over 12	Over 14
E-9	0.00	0.00	0.00	0.00	0.00	0.00	2496.90	2552.70	2610.60
E-8	0.00	0.00	0.00	0.00	0.00	2093.70	2153.70	2210.40	2267.70
E-7	1461.60	1578.00	1636.20	1693.80	1751.40	1807.20	1865.10	1923.30	2010.30
E-6	1257.60	1370.70	1427.70	1488.60	1544.40	1599.90	1658.70	1744.20	1798.80
E-5	1103.40	1201.20	1259.70	1314.30	1401.00	1458.00	1515.60	1571.40	1599.90
E-4	1029.30	1087.20	1151.10	1239.90	1288.80	1288.80	1288.80	1288.80	1288.80
E-3	969.90	1023.00	1063.80	1105.80	1105.80	1105.80	1105.80	1105.80	1105.80
E-2	933.30	933.30	933.30	933.30	933.30	933.30	933.30	933.30	933.30
E-1	832.80	832.80	832.80	832.80	832.80	832.80	832.80	832.80	832.80

SOURCE: MSG Gary L. Smith (Ret.) and Debra M. Gordon, *The Uniformed Services Almanac,* Falls Church, Va.: Uniformed Services Almanac, Inc., 1994.

NOTE: For the senior member of the service, Basic Pay is $3,906.90 per month, regardless of years of service. The complete table reports rates beyond the "Over 14" Years of Service category.

could be viewed as compensating wage differentials[2] that make up the differences in risks, transferability to the civilian sector, and other nonpecuniary traits that vary across military occupations. Because of the wide range of eligibility for Special and Incentive Pays and their varying amounts, this component is likely to vary greatly across service members, as well as across years of service.

Enlistment/Reenlistment Bonuses

We include in this category both Enlistment Bonuses and Reenlistment Bonuses. The Enlistment Bonus may be offered as an additional accession inducement to new recruits in military skill specialties designated "critical." Service members qualify for the Reenlistment Bonus if they reenlist immediately so that their service is continuous. This bonus is designed to ensure adequate levels of experienced personnel in the services.

Basic Allowance for Quarters (BAQ), or Housing

The components of compensation described thus far are all pays that would be received through enlistees' monthly paychecks. But the military also provides in-kind transfers of services (or a cash allowance) to service members. Military compensation for service members' housing is one such component. Enlistees either live in base housing or off base. If they live off base, they are eligible to receive BAQ.[3] Depending on their assignment location, members may also receive a Variable Housing Allowance (VHA), which was designed to supplement BAQ for members living in areas of the United States having high housing costs. Members living overseas are eligible for the Overseas Housing Allowance (OHA), which is a per diem to compensate members for additional housing expenses associated with living overseas.

Military compensation for service members' housing—including base housing, BAQ, VHA, and OHA—is based on characteristics that are presumably correlated with productivity, such as tenure and rank, and with whether the service member has dependents.

---

[2]A *compensating wage differential* is the extra amount an individual would have to be paid to be induced to accept an undesirable job attribute such as risk or heat. See Rosen (1986) for a complete discussion.

[3]Note that members without dependents who live in base housing are eligible to receive partial BAQ, a substantially smaller BAQ payment. Partial BAQ was initiated in FY77 to rectify a redistribution in the annual pay raise (see U.S. Department of Defense, 1991). Partial BAQ is $6.90 per month for an E-1 and $18.60 for an E-9.

The amount and type of housing compensation thus depend on the individual's rank, whether the individual lives on base, the cost of living in the stationed area, and whether the person lives overseas. Note that while the cash transfer to service members living in base housing is zero, some benefit is still associated with the in-kind housing provided the individual. We discuss the valuation of this benefit in Section 3.

Basic Allowance for Subsistence (BAS), or Board

Another component of compensation that may be allocated either by a cash payment or an in-kind transfer is BAS, or board. BAS is granted when in-kind rations (i.e., food) are not available to the service member or when the member has been granted permission to take meals outside the military dining facility. BAS-payment amounts depend on officer or enlisted status. Unlike BAQ, enlisted BAS does not depend on the presence of dependents. A variety of references publish BAQ and BAS schedules similar to the Basic Pay Table presented earlier (for example, Smith and Gordon, 1994).

Medical Benefits

With rapidly escalating health care costs, Medical Benefits are becoming an increasingly important component of military pay. Health care benefits to members and their dependents are paid through an in-kind transfer of both medical services and private insurance.[4] These in-kind transfers are delivered through a system of 500 military treatment facilities (MTFs) in the United States, operated by DoD. Members and their dependents receive treatment at MTFs free of charge. In addition, dependents of active-duty personnel are eligible for insurance through the Civilian Health and Medical Program of the Uniformed Services (CHAMPUS), a traditional fee-for-service health insurance program. This insurance provides comprehensive coverage for care sought in the private sector.

All service members receive these medical benefits, but not all service members have the same number of dependents. With private-sector insurance plans costing more than $5,000 for family coverage, military health-care coverage is an

---

[4]Retirees and their dependents are also eligible to receive military treatment facility (MTF) services and Civilian Health and Medical Program of the Uniformed Services (CHAMPUS) benefits free of any premium charge. This added benefit will be treated simultaneously with Retirement Benefits, discussed in Section 3.

important component of total compensation; in addition, its value will change according to the number of family members covered.

Retirement Benefit

Our final component of military compensation is the Retirement Benefit program. In addition to the goal of providing a reasonable income upon retirement, which the military system shares with most civilian plans, the military retirement system has the additional mission of helping maintain the desired age and promotion structure in the military. This additional mission skews the system's design toward inducing personnel to separate at relatively young ages.

Rather than providing benefits starting at an older age, such as at 65 with civilian plans, the military plan provides retirement benefits to individuals with 20 years of service (at which point most enlistees separate). Unlike most civilian plans, military personnel are not vested in their plan until they have completed all 20 years. Therefore, whereas the promise of future streams of retirement benefits is a small fraction of the value of military compensation in the early years of service, this future value becomes a large fraction of the value of compensation— and a large inducement to stay in the military—in later years of service.[5]

Components Excluded from the Study

We do not attempt to measure several other components of military compensation in this study: paid vacations, the use of military commissaries, life-insurance plans, educational benefits,[6] and a host of perquisites known as morale, welfare, and recreation (MWR) benefits, which are designed to enhance social and recreational opportunities. As with military service, most occupations have both positive attributes, which could raise the value of the job to employees, and negative attributes, which could lower it. We do not include these components in our study of compensation, because we do not have good measures of them. And, in most compensation studies, the value of perquisites and the positive amenities of occupations are not considered as portions of compensation per se; rather, they are characteristics of the job that, all else being equal, could lead employees to be willing to trade off wages for the positive or negative job attributes.

---

[5]See Asch and Warner (1994) for a more complete discussion of the retirement system.

[6]Asch and Dertouzos (1994), for example, estimate that the actuarial value of Army College Fund benefits at the time of enlistment is $1,144 per recruit who qualifies for the program.

We also do not include VHA and OHA in our measures of compensation. Designed to equalize the housing allowance afforded members in areas with different costs of housing, VHA and OHA do not add value to members' compensation; instead, they ensure that everyone receives an equal real value of the BAQ. We are interested in measuring real rather than current values of compensation. To do so, we also adjust values in different years to reflect the rise in the cost of living over time, expressing all values in 1992 dollars.

We also do not attempt to measure the value of on-the-job training or other forms of training. Other studies have shown that veterans may receive a benefit from prior military service in the form of higher civilian wages than nonveterans and that this premium is most likely from training received in the service (see, for example, Angrist, 1989, 1990; Berger and Hirsch, 1983; Rosen and Taubman, 1982).

Finally, we do not account for the fact that personnel living on base do not have to pay for utilities. Doing so would make the relative value of on-base housing higher and would raise the variance in earnings due to In-Kind Housing.

Summary

In summary, military compensation is a bundle of components that vary in their incidence of receipt and in the amount personnel receive. Hence, in our investigation of compensation variation, we explore differences in benefit receipt and benefit amounts. The next section indicates how we measured benefit receipt and amounts.

3. Valuing Military Compensation

In this section, we describe how we placed a value on the components of compensation discussed in the preceding section and how we valued an individual's military-compensation package. For components of pay that are issued as cash payments, including Basic Pay, BAQ, BAS, Special Pays, and Enlistment/Reenlistment Bonuses, the value of the component is simply equal to the payment. However, many of the components, such as On-Base Housing, Medical Benefits, and the future Retirement Benefit, are not issued as cash payments. We convert our measures of these benefits to a cash value and add them to the cash payments to arrive at compensation measures.

We first describe our data set. Then we explain how we valued cash and noncash components of compensation. Finally, we define the measures of compensation that we analyzed.

Data Set

Our measures of both cash and noncash components of compensation are from a longitudinal data set created by the Defense Manpower Data Center (DMDC) especially for this project. The data contain demographic characteristics of individuals, along with compensation information for every year the service member is in the military. To reduce sample size, we selected subsamples of every other entering enlisted cohort, starting with the cohort that began service in 1978 and ending with the 1990 cohort. Selecting every other cohort still generated enough observations to allow us to conduct statistical tests that examined differences among subpopulations.

Our data set contains matched records from four different DMDC data sources: the Military Entrance Processing Command (MEPCOM) file, Active Duty Master files, the Active Duty Military Pay Files (the "JUMPS" files), and Active Duty Loss files. The MEPCOM files contain one record per person and report accession information and static demographic variables, such as AFQT score, race, sex, and service. The Active Duty Master files are collected each year a member is on active duty and include annual demographic information, such as marital status, number of dependents, and education. The annual-pay variables and pay grade come from the JUMPS files, which begin in 1985 and end in 1993; and separation information is reported in the Active Duty Loss files. Each record for an individual contains the MEPCOM section, followed by Active Duty Master records for each year of enlisted service, JUMPS files for 1985–1993, and a Loss section when the service member separates. Figure 3.1 illustrates the structure of each record. All the data are reported as of the end of each fiscal year.

To measure earnings and the factors contributing to earnings, we used primarily the combined Master and JUMPS records, which exist only from 1985 to 1993. The JUMPS records we have report the service member's compensation from each pay (for example, Basic Pay, BAQ) in the last month of each fiscal year. Because our data include individuals from cohorts accessing between 1978 and 1990, we observed service members with years of service ranging from as few as one to as many as 16. Table 3.1 reports the cohort and year-of-service distribution for our 1985–1993 analysis file. Appendix Table A.1 reports the number of individuals from each cohort observed at each year of service.

We sampled groups other than white males at higher rates to ensure large enough sample sizes at later years of service to conduct statistically significant

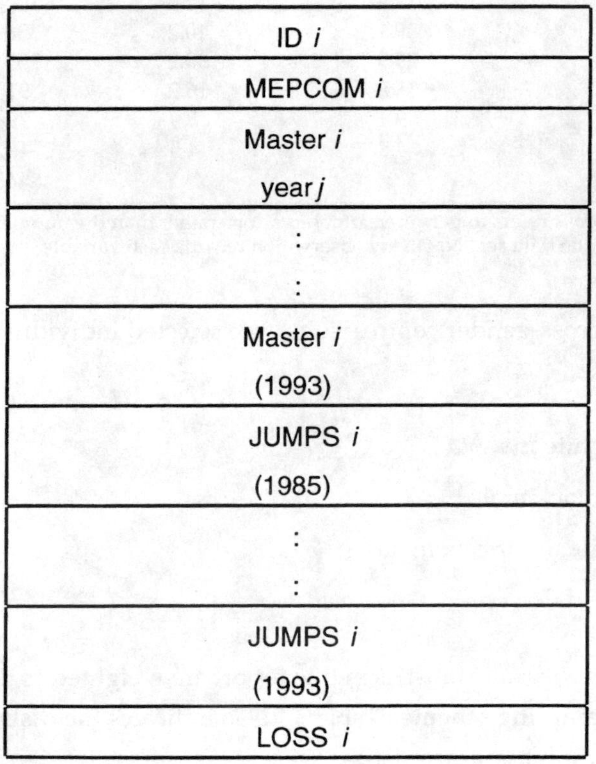

ID i
MEPCOM i
Master i year j
:   :
Master i (1993)
JUMPS i (1985)
:   :
JUMPS i (1993)
LOSS i

**Figure 3.1—Structure of the Data Set: Layout of a Complete
Record for Individual i Who Entered in Year j**

Table 3.1

Percentage of Sample and Number of Observations by Characteristic for 1985–1993 Analysis File

Characteristic	Unweighted Percentage of Sample, All Cohorts	Weighted Percentage of Sample, All Cohorts	Number of Observations,[a] All Cohorts
Female	30.1	11.7	254,789
White	51.9	71.5	439,136
Black	34.8	23.3	294,808
Other Race	13.3	5.2	112,979
Army	37.0	35.4	313,023
Navy	26.4	26.5	233,652
Marines	11.6	13.0	98,526
Air Force	25.0	25.1	211,772
1978 Cohort	8.4	8.0	71,289
1980 Cohort	11.0	10.0	92,805
1982 Cohort	11.9	12.1	100,783
1984 Cohort	19.0	19.5	160,825
1986 Cohort	20.0	20.8	169,607
1988 Cohort	17.3	17.3	146,682
1990 Cohort	12.4	12.3	104,982
YOS 1–3	39.1	40.2	330,808
YOS 4–6	29.9	30.2	253,550
YOS 7–9	16.8	16.2	91,497
YOS 10–12	9.4	8.9	79,735
YOS 13–17	4.8	4.5	40,270
Total			846,973

[a]*Observations* refers to person-year observations rather than the number of individuals in the data set. Not every observation records each variable.

across-race and across-gender comparisons. We selected individuals by race and gender as follows:

- Every tenth white male
- Every fourth black male
- Every nonwhite, nonblack male
- Every other female.

Unless noted otherwise, all statistics in this report are weighted to account for this nonrandom sampling scheme. Table 3.1 also indicates the distribution of our sample by demographic characteristics.

It was essential that we use a panel data set that follows the same individuals over time rather than repeated cross-sectional data that report information on individuals in each calendar year without matching information on the same individuals across years. Only with panel data could we impute values for the Retirement Benefit we need to be able to estimate expected career profiles as a function of individual characteristics. This calculation requires data on the same individuals progressing through the service. We discuss the use of this longitudinal information when we describe how we estimated the Retirement Benefit (Appendix D).

Valuing Cash Components

Monthly compensation data are reported in the JUMPS file for the last month of the fiscal year. Cash transfers to each service member are reported for over 20 different pay categories. As indicated in Appendix Table B.1, we collapsed these pay categories into the following components:

- Basic Pay
- Special Pays
- Enlistment/Reenlistment Bonuses
- BAS
- BAQ.

Hence, for every service member in our sample, we have the monthly cash amount received in the final month of the fiscal year for every year that individual was enlisted.

During the period our data cover, Enlistment/Reenlistment Bonuses may have been paid in lump-sum amounts, over a number of months, or in a combination of the two. Our data report the payments that a service member received in the last month of the fiscal year, which therefore include a combination of Enlistment/Reenlistment Bonuses paid as lump sums and in monthly amounts. Since our unit of analysis is monthly income, we did not adjust for variation in the way these bonuses are paid.

Estimating Noncash Components

Those components of compensation that are not allocated to service members as a cash transfer include the following:

- On-Base Housing
- Tax Advantage

- Medical Benefits
- Retirement Benefit.

We converted each of these transfers into a cash value so that they might be combined with the cash components to obtain a compensation measure.

An entire body of literature describes ways to place a cash value on in-kind transfers (see, for example, Moffitt, 1989; Smeeding, 1977; Murray, 1983). The analyses in this literature are based on standard microeconomic theory, which posits that individuals will value an in-kind transfer as being less than or equal to an equal-dollar cash transfer. Hence, the literature on in-kind transfers often determines the cash amount that individuals would have to receive in order to be indifferent between receiving the transfer in kind or as cash. That is, this literature identifies the *worth*—the utility-equivalent cash value—individuals place on the in-kind transfers they receive rather than measuring the actual allocation of the in-kind transfers.

While determining the utility-equivalent cash value of in-kind benefits is a valid issue for further study, it is beyond the scope of this study. In this study, we are less interested in determining the equivalent cash value that service members would be willing to accept in place of their in-kind benefits than in determining the actual allocation of the dollar value of benefits among service members. That is, for our purposes, what is needed is an estimate of the dollar value of benefits members are receiving and how this value varies. One way to estimate this value would be to determine how much it costs the military to provide each of these benefits. Since we do not have data on these costs and they would be very difficult to obtain, we use alternative methods of imputing values. We now describe how we arrived at the dollar value of each of the in-kind benefits.

On-Base Housing

We had to determine a method for calculating a dollar value of the In-Kind Housing transfers. As discussed earlier, service members and their families who live on base receive housing at no cost to them; those who live off base receive BAQ, plus VHA if they live in a high-cost area. Also, members who live on base and have no dependents receive a partial BAQ, which is less than one-tenth of full BAQ.

Our data set indicates whether individuals received BAQ or VHA, the size of the BAQ[1] or VHA payment, their unit, and the number of family members in the household. We assumed that service members live in base housing if they meet at least one of the following two conditions: (1) They receive no BAQ or VHA, or (2) they receive BAQ the size of partial BAQ. We also assumed that individuals live on the base to which their units are attached when we observed them.

Each calendar year in its *Census of Uniformed Service Members* (now called *Variable Housing Allowance Housing Survey*), DoD collects information about rental-housing payments made by service members living off base. From this census, for each pay grade, number of dependents, and base combination, DoD computes a median monthly rent paid. We used this median value as an estimate of the amount an enlisted person with those same characteristics would pay for housing in his or her base location. That is, for a service member with attributes i (pay grade, number of dependents), living on base j, at time t, we approximated the value of on-base housing with R_{ijt}, the median monthly rent paid by members with the same characteristics ijt. DoD data on median housing values were available only for fiscal years 1991, 1992, and 1993.[2] For this reason, we were able to estimate the housing value only for those three fiscal years.[3]

Why not just use BAQ plus VHA as the value for on-base housing? Two pieces of evidence suggest that this is not an appropriate estimate of the value of base housing. First, anecdotal evidence indicates that while space is often available for single members in barracks, queues frequently exist for base housing for members with families. In other words, given the available housing subsidy—BAQ plus VHA—at least some members prefer on-base housing to off-base rental units. If service members faced a choice between obtaining base housing and receiving a payment of the true market value for the housing, they should be exactly indifferent between living in on-base housing and receiving the subsidy. Thus, the existence of queues implies that BAQ plus VHA underestimates the value of on-base family housing.[4]

Because the subsidies do not fully cover the median rent, the existence of these queues is not surprising. The total cash housing subsidy given members living off base is BAQ plus VHA. BAQ is set at 65 percent of the national median rental

_____

[1]We included both full and partial BAQs in our measure of BAQ.

[2]DMDC provided us the base locations of the units of individuals in our sample. We then matched the appropriate R_{ijt} to each member of our sample.

[3]Note that this analysis assumes that those who choose on-base housing do not differ from those who live in off-base housing in unobservable ways, such as distance to spouse's employment or a desire to own horses or other livestock, that would affect the rental rate.

[4]This argument does not apply to single service members, since there is usually no queuing for space in the barracks.

cost for a service member in a particular pay grade and dependent status, or $BAQ = .65R_{it}$. VHA varies across bases, pay grade, and whether the service member has dependents. VHA is computed as the difference between the average monthly housing costs for members in a particular pay grade (with and without dependents) at a particular base, and 80 percent of the national median housing value for individuals in each pay grade. In other words,

$$VHA_{ijt} = R_{ijt} - .8R_{it}, \tag{3.1}$$

where VHA_{ijt} is the VHA for a service member with attributes i (pay grade, number of dependents) living on base j, at time t. Equation (3.2) shows that the housing subsidy is less than the median rental cost for any given family size and base location:

$$BAQ_{it} + VHA_{ijt} = .65R_{it} + (R_{ijt} - .8R_{it}) = R_{ijt} - .15R_{it} < R_{ijt}. \tag{3.2}$$

Thus, on the average, members who rent off base must use some of their own money to make up the difference between their housing subsidy and rent; members who receive on-base housing do not have to use any of their own funds. This is the second piece of evidence suggesting that BAQ plus VHA is a poor measure of the value of base housing. In fact, military housing subsidies were *designed* to cover less than 100 percent of off-base housing costs (see U.S. DoD, 1991).

Note that our measure of On-Base Housing does not control for variation in the price of housing across base locations. If a price index of military housing locations were available, we could adjust for these differences. We were not able to obtain such a price index. Our estimates of the value of On-Base Housing are likely to overstate the value in high-cost areas and yield larger estimates of the variance in the value of On-Base Housing.

Tax Advantage

We used the DoD actuarial method of computing the federal tax advantage that results from the nontaxability of many military pays and benefits. This method calculates the additional pre-tax amount an individual would have to be given to be indifferent between the benefit's being taxable and nontaxable. The calculation allows for the possibility that the additional pre-tax payment might bump the individual up into the next tax bracket. While not making a large difference in the size of the Tax Advantage, this bump-up affects approximately 4 percent of service members. We describe these methods in detail in Appendix C.

Medical Benefits

Service members and their dependents are eligible for free medical services through the system of MTFs. Dependents are also eligible for CHAMPUS benefits, as described in Section 2. Thus, the military provides service members and their dependents with a comprehensive set of benefits. In fact, as Table 3.2 demonstrates, CHAMPUS insurance provides a comprehensive set of benefits similar to those of typical private plans found in medium-sized and large private firms. Thus, we approximate the value of CHAMPUS benefits using the average premiums for employer-provided plans.

On average, the annual health insurance premium for a single worker costs $2,147, and a family policy costs $5,292.[5] Thus, we assumed that insurance for a service member with no dependents costs $2,147. Similarly, we assumed that the cost for a service member with one dependent is $4,294, and the cost for a service

Table 3.2

Comparison of CHAMPUS and Private Insurance

	CHAMPUS Active-Duty	CHAMPUS Retired	Private FFS[a]
Deductible	$50	$100	$100
Physician Services			
In hospital	Covered in full	Covered in full	Covered in full
Office visits	Ben.[b] pays 20%	Ben. pays 25%	Ben. pays 20%
Outpatient Mental Health	Ben. pays 20%	Ben. pays 25%	Ben. pays 50%
Surgical Coverage			
Inpatient	Covered in full	Covered in full	Ben. pays 20%
Outpatient	Ben. pays 20%	Ben. pays 25%	Ben. pays 20%
Hospital Coverage	Ben. pays $8/day	Ben. pays $210/day up to 25% of charges	Covered in full or Ben. pays 20% of semiprivate rate[c]
Prescription Drugs	Ben. pays 20%	Ben. pays 25%	Covered in full

[a]Private fee-for-service (FFS) consists of the modal (most-frequent) benefit provided in medium-sized and large firms (Bureau of Labor Statistics, 1994).

[b]There are two modal benefits in this case.

[c]Ben. stands for *beneficiary*.

---

[5]Average premiums come from a national survey of employers done by the Health Insurance Association of America in 1991. The premiums were inflated to 1994 using an annual inflation rate of 7.5 percent, which corresponds closely to the medical consumer price index (CPI) for those years.

18

member with two or more dependents is $5,292.[6] We used the following formula for the contemporaneous value of military-provided health insurance:[7]

$$V(\text{ANNUAL BENEFIT}) = \$2,147 + (\$2,147 \times \text{FIRST DEPENDENT}) + \$1,042 \times I$$

$$\text{where } I = \begin{cases} 0 \text{ if there is one dependent} \\ 1 \text{ if there is more than one dependent.} \end{cases} \tag{3.3}$$

Retirement Benefit

Military personnel are not eligible to receive retirement benefits until they have served 20 years, and benefit receipt is independent of age. Despite the fact that no retirement benefits are received while the member is enlisted, it is important to include the expected value of the future benefits, for two reasons. First, previous work has shown that the retirement benefits of military personnel are a substantial fraction of their lifetime wealth (Phillips and Wise, 1987). Second, the promise of the military retirement benefit after 20 years of service is a powerful inducement for individuals to remain in the service after having served 10 or more years (Phillips and Wise, 1987; Asch and Warner, 1994).

The key issue in valuing the retirement benefit is the following: What value does a person receive before 20 years of service from a retirement plan that vests him or her at 20 years of service? Both a behavioral approach and an actuarial approach can be taken to measure the value of retirement benefits. The behavioral approach identifies the cash payment that would make a person indifferent between receiving the cash and receiving the addition to expected retirement benefits granted by the additional year of service (see the models in Black, Moffitt, and Warner, 1990; Gotz and McCall, 1983). As discussed above, we are less interested in converting a year's retirement benefits into a utility-equivalent cash amount than in studying the way the dollar-benefit accrual varies across individuals with different characteristics.

We took a more actuarial approach, viewing the retirement benefit as a stock available after 20 years of service, but expecting the value of the stock to change in each year of service because the individual has one less year of discounting the benefits, has an updated life expectancy, has a higher probability of realizing retirement, and has a different expectation of future benefits because he or she has completed one more year of service.

---

[6]Most firms determine copayments by whether the employee has zero, one, or two or more dependents. For service members with one dependent, that dependent is typically their spouse.

[7]Military benefits also accrue upon retirement. We value the retiree benefit separately as part of our strategy for valuing pension benefits (see Appendix D).

Our measure of the retirement benefits a person gets in each year is a *flow*: an estimate of the change to the net present discounted value that each year of service yields, conditional on individual characteristics. The method we used is described in detail in Appendix D.

Measures of Compensation

In this subsection, we describe the ways in which we bundled the components just described to form measures of compensation. Not all members receive all components, and the components' values differ across service members—inducing variation in compensation. More important, we could not infer the variations by looking at pay tables and other compensation formulae in isolation. We employed four different compensation measures to describe military pay: Basic Pay, Regular Military Compensation (RMC), Cash Compensation (Including and Without In-Kind Housing), and Total Compensation. We define them more precisely below: Each measure provides different insights into the structure of military pay.

Basic Pay

We examined Basic Pay in isolation from the other components for several reasons. First, Basic Pay constitutes the largest and most visible portion of a service member's compensation. Second, Basic Pay is completely determined by an individual's pay grade and years of service; thus, it is perceived as being the least-flexible component of compensation.

Regular Military Compensation

Regular Military Compensation, or RMC, is a measure of military compensation that is commonly used to evaluate the take-home pay of military personnel. RMC includes Basic Pay, BAQ, BAS, and the Tax Advantage. No in-kind transfers are included, nor are cash payments other than BAQ and BAS included. Since this measure is so widely used in studies of military compensation, it will facilitate the comparison of our results with those of other studies.

Cash Compensation (Including Housing and Without Housing)

We also constructed a measure of cash compensation that includes all components of compensation that might appear in a service member's paycheck. Cash Compensation includes all (potential) cash payments, including Basic Pay,

BAQ, BAS, and Special Pays. This measure is easily identified by the service member, and is also easily measured. Further, Cash Compensation is not subject to variations in utility in the same way that in-kind transfers might be, and hence does not derive its value from the (heterogeneous) preferences of individuals.

Because Cash Compensation may or may not include BAQ and because BAQ is such a large fraction of Cash Compensation, we examined two Cash Compensation measures: One is purely Cash Compensation; the second is all Cash Compensation plus the estimated value of In-Kind Housing for those not receiving BAQ. No in-kind transfers other than the housing value are included in measures of Cash Compensation.

Total Compensation

Our most comprehensive measure is Total Compensation. It encompasses both Cash Compensation and noncash components (i.e., On-Base Housing, Tax Advantage, Medical Benefits, and the Retirement Benefit). Total Compensation comes closest to capturing the entire benefit package a service member receives. This measure displays the most variation across service-member characteristics.

Recall that, because of data limitations, we could estimate the housing value for only fiscal years 1991–1993. Since Housing is a component of Total Compensation and Cash Compensation Including Housing, we used only data from those fiscal years to create these two compensation measures. When discussing the results that follow, we refer to the complete set of data as "the full sample" and the restricted sample for fiscal years 1991–1993 as "the housing sample."

Summary

Table 3.3 summarizes the components that make up each compensation measure. In Section 4, we explain the conceptual framework that guided our analysis and describe the empirical methods we used.

Table 3.3

Summary of Components in Each Compensation Measure

Components	Basic Pay	RMC	Cash Compensation Without Housing	Cash Compensation Including Housing	Total Compensation
Basic Pay	X	X	X	X	X
BAQ		X	X	X	X
BAS		X	X	X	X
Housing Benefit				X	X
Special Pay			X	X	X
Enlistment/ Reenlistment Bonus			X	X	X
Medical Benefits					X
Tax Advantage		X			X
Retirement Benefit					X

4. Methods

Conceptual Framework

Recall that our goals were to document how enlisted personnel are paid and to measure the patterns of enlisted compensation across service-member characteristics. We explain the way we approached this problem with the help of Figure 4.1.

Individual characteristics are related to whether an individual receives the various components of compensation and to the amount of each component received. The left panel of Figure 4.1 shows the characteristics we examined in this study, and the arrow indicates that these characteristics are related to the compensation components we studied, listed in the second panel. Whether a service member receives various components of compensation and the size of the component yields different values for the compensation measures we examined. This relationship is illustrated by the second arrow pointing from the components to the compensation measures. This scheme indicates how individual characteristics are translated into our compensation measures.

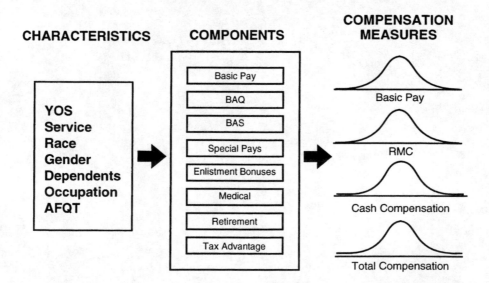

Figure 4.1—Conceptual Framework

Our empirical analysis reflect the conceptual scheme in Figure 4.1. First, since the compensation measures comprise multiple components, we examined the amount each component contributes to aggregate pay. For example, is aggregate pay mostly determined by just a couple of components, or do all the components make equal contributions to the level of pay? Second, we investigated the degree to which the receipt and size of the components depends on individual characteristics. Finally, we investigated the relationship between individual characteristics and the compensation measures. Figure 4.2 summarizes the three relationships we examined.

Estimation Strategy

For the first of the three relationships, we began by determining how the components affect the average compensation-measure *levels*, then we examined each component's contribution to the total *variation* in the compensation measures. To understand the importance of this distinction, consider the contribution of Basic Pay to our measure "Cash Compensation." Below, we show that Basic Pay accounts for about 80 percent of Cash Compensation—that is, on average, about 80 percent of Cash Compensation is Basic Pay. While Basic Pay contributes a substantial amount to the *level* of Cash Compensation, it may not contribute much to the *variation* in Cash Compensation.

In analyzing the variation, we were interested in how much of the difference in the Cash Compensation that individuals receive can be attributed to Basic Pay. As reported below, in contrast to its importance in determining the level of Cash Compensation, Basic Pay makes a small contribution to the total variation in Cash Compensation. Put another way, suppose all service members had exactly the same Basic Pay, but other cash compensation varied. In this case, while Basic Pay may constitute a large fraction of Cash Compensation, it would explain none of the differences in pay across individuals.

To identify the contribution of components to the *level* of the compensation measures, we simply calculated the average fraction of the outcome variable

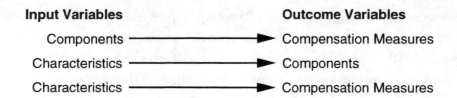

Figure 4.2—Three Relationships We Examined

accounted for by the input variable. The average contribution of a component a to compensation measure c equals the mean of a/c.

To identify the contribution of components to the variance in the compensation measure, we show the percentage of variance in the compensation measures that is due to the variance in each component. The variance (var) of the sum of variables is the sum of their individual variances plus the sum of the covariance (cov) terms between the variables:

$$\text{If } c = a + b, \text{ then } \text{var}(c) = \text{var}(a) + \text{var}(b) + \text{cov}(a,b). \tag{4.1}$$

We report the fraction of total variance in the compensation measure from each component's individual variance, along with the fraction of total variance made up of the summed covariance terms. Continuing the same example, we report the mean var (a)/var (c), var (b)/var (c), and cov (a,b)/var (c).

That we imputed values for some components of compensation is likely to influence our summary measures of the variance contributed by those components. For example, because we assigned the value of On-Base Housing as the median housing cost of off-base housing at each base location (for a particular pay grade and family structure), the variance contributed by the Housing component will be smaller than the true variance. The estimated variance of the compensation measure containing that component will also be smaller than the true variance. However, the ratio of the two is likely to be smaller than the true variance, because the reduction in variance caused by assigning the average value is likely to be a larger fraction of the variance of the component than of the variance of the compensation measure, which is the sum of several components.

Examining the relationship between the components and the compensation measures answers the question of how service members are compensated rather than how much they are compensated. These results indicate the degree to which service members are remunerated in the form of in-kind benefits as opposed to cash payments, deferred benefits as opposed to contemporaneous benefits, and components that are the same for all members instead of varying by member characteristics.

For the second and third of the three relationships in Figure 4.2, the contribution of service members' characteristics to the *level* of components and measures of compensation they receive, we used regression analysis to answer the question of how much compensation service members with different characteristics receive. The regression results indicate the ways in which the components and compensation measures vary with an individual characteristic, holding all other measured characteristics constant.

This approach differs from a univariate description of variation in compensation by each characteristic, in which it would not be possible to distinguish between the possibility that individuals with a particular characteristic received more compensation and the possibility that individuals with a particular characteristic were also more likely to exhibit another characteristic, and that individuals with the second characteristic receive more compensation. For example, if we examined the compensation of females without controlling for other characteristics, we would not be able to evaluate whether any differences in male and female compensation were related to being male or female per se, or merely related to the fact that women were more likely to be observed in particular occupations and branches of service, and thus that the differences between male and female compensation were actually due to differences in occupation and service by gender. By using regression analysis, we accounted for other observed characteristics to identify the variation in compensation due to gender while holding constant the other characteristics, such as service and occupation.

The regression model takes the form of Equation (4.2):

$$y = x_1\beta_1 + x_2\beta_2 + \ldots + x_n\beta_n + \varepsilon, \tag{4.2}$$

where y is the component or measure of compensation, the x_i are the service-member characteristics believed to influence compensation, and the estimated β_i indicate the unit change in y due to a unit change in x_i. To summarize, the regressions of the components and compensation measures on service-member characteristics will document how compensation varies by these characteristics.

Researchers who estimate civilian-sector wages typically use the log of earnings as the dependent variable, whereas we used the un-logged earnings as the dependent variable in our regressions. The justification for their using the log transformation of earnings is that earnings are usually not distributed normally. Earnings tend be left skewed—that is, the distribution is not symmetric, and the upper tail of the distribution spans a wide range of high values—and they are better described by a log-normal distribution. Also, the coefficient estimates in the log-earnings case are interpreted as a percentage change rather than a unit change, as in the un-logged case.

We did not find that most of the military-pay variables we studied had a log-normal distribution, however. We could not reject the hypothesis that the distributions were normal for Basic Pay and most of the components. Although the distributions of Cash Compensation and Total Compensation were not normal, we chose to estimate these measures using the same specification as that used for the Basic Pay measure and the components to facilitate comparisons of the results. In addition, no theoretical justification led us to believe that characteristics should induce a percentage change rather than a unit change in the outcome variables.

Since we observed multiple wages for individuals over time, we could have included an individual-specific effect in our wage functions to account for unobserved individual-specific factors—such as unobserved ability or motivation—that contribute to wages (see Amemiya, 1985, or Greene, 1993). Two common approaches to accounting for the individual-specific effect are using fixed effects or random effects. The fixed-effects approach uses only time-varying characteristics to identify the individual-specific effect. Since some of the primary characteristics of interest in our model are not time-varying, the fixed-effects approach is not a good strategy for this analysis. The random-effects model assumes that the individual-specific effect is uncorrelated with the regressors. Since we included a measure of ability—AFQT score—this assumption is likely to be violated. As a result, we did not incorporate an individual-specific effect in our earnings estimates.[1]

Descriptive Interpretation

This analysis was meant to document patterns of enlisted compensation rather than to explain why the patterns exist. Our regression estimates indicate the reduced-form (i.e., net) relationships between service-member characteristics and compensation outcomes. For instance, if we had wanted to understand the association between Basic Pay and individual characteristics, we could have modeled the relationship between promotion speed, years of service, and individual characteristics. Then, we could have used these results, along with the Basic Pay Table, to relate the characteristics to Basic Pay. Instead, we took a reduced-form approach that indicates the relationship between Basic Pay and characteristics, without the intermediate step of modeling promotions. We took a similar approach for other pay outcomes.

It is important to keep this in mind when interpreting the estimates. For instance, the amount of a compensation component a member receives is related to whether that individual is eligible to receive the pay at all and, if so, the amount that he or she receives.[2] We did not model both receipt of the pay and

---

[1]Note, however, that because we have repeated wage observations on the same individuals, our estimated standard errors may be too small (see Greene, 1993). We reestimated some of our equations in software that would account for repeated observations (STATA), using a random-effects estimator. We found that the significance of variables changed only rarely and that the coefficient estimates were not substantively different from our ordinary least squares (OLS) regression estimates.

[2]In other words, observations containing dependent-variable values of zero are included in the regressions.

the amount of pay, given receipt; rather we estimated a reduced-form model that represents the net effect of both incidence and amount.

In addition, while incidence and receipt of many of the pay components depend on promotion speed, changes in marital status, and/or other behavioral factors, we did not attempt to model the behavior involved in these inputs into compensation outcomes.

Similarly, our estimates indicate patterns of variation in compensation rather than necessarily measuring productivity differences or behavioral differences across characteristics, as would a wage function. We included many of the same explanatory variables that appear in civilian wage-function estimates; however, our results do not share the same interpretation with the wage-function result. Coefficient estimates in wage functions presumably represent the productivity of various individual characteristics, such as experience, schooling, or test scores (see, for example, Becker, 1975; Mincer, 1974; Lillard, 1977; Willis and Rosen, 1979; O'Neill, 1990).

The application of the wage-function analogy for the military sector is likely to be confounded by a number of factors, including the fact that the military is a strict internal labor market (see Rosen, 1992, for a discussion), that continuation in the military yields an option value to receive future military income streams, and that the design of retirement benefits after 20 years of service is likely to divorce current productivity from current compensation (see Lazear, 1979; Asch and Warner, 1994). Hence, a complete understanding of military-compensation patterns would necessarily model separation, retention, promotion, retirement, individual choices regarding family formation, and more.[3] Our baseline description of compensation variation across individual characteristics helps identify the modeling exercises that would most likely yield the most-useful insights into military compensation.

In our military-earnings regressions, we included a set of individual characteristics that are measured in our data set and that would be expected to yield different earnings patterns either as a result of what is known about the

---

[3]Note that we include years of service as a covariate in our regressions. As with the civilian-wage literature on tenure (see Topel, 1991), we did not expect that individuals with higher years of service are a random subset of all individuals who enter military service. As in any job, accumulating a large number of years of service is the result of retention decisions, accumulation of job-specific capital, and other factors. This does not pose a problem for the interpretation of the results, however. We did not seek to estimate what a random service member would earn given some number of years of service; rather, we sought to describe what service members earn who actually attain that number of years of service. Hence, the earnings of the selected subset of individuals who remain after each year of service was precisely what we wanted to measure.

allocation of military pays or because it is expected that age-earnings profiles in the military would differ on these characteristics.[4]

Characteristics might be related to pay outcomes for a number of reasons. First, pay schedules might explicitly allow for pay differences on some characteristics, such as years of service, number of dependents, or occupation. Second, some characteristics might be related to productivity, as might AFQT score or years of service. Third, characteristics might be associated with differences in compensation due to selection. An example of this association might be marital status: Some of the literature on civilian earnings hypothesizes that married individuals earn more in the civilian sector because there is positive selection on unobservable (to the econometrician) productive characteristics in the marriage market (see summary in Daniel, 1994). Finally, there may be other sources of correlation, such as occupational-assignment rules affecting women's compensation or the supply of barracks slots influencing the housing benefits for single people.

Appendix E defines all the individual characteristics we included in our regression equations. The next section presents the results of our empirical analysis.

---

[4]We did not examine the variation in compensation by education in these estimates, because there was little variation in education in our data.

5. Results

In this section, we present results on the effect of the input variable on the outcome variable for each of the three relationships described above: the contribution of components to the measures of compensation, the contribution of individual characteristics to receipt of the components, and the contribution of these characteristics to the measures of compensation. For each of these three relationships, we report results separately by service, since the results vary significantly by service. We also examined the possibility of differences in estimates by gender, race, and other variables, but running separate equations by gender, race, or other variables did not result in substantively different conclusions.

First, we present the mean and standard deviation of our variables by service:[1] Table 5.1a and Table 5.1b include the characteristics; Table 5.2a and Table 5.2b, the components; and Table 5.3a and Table 5.3b, the compensation measures. For each table, set *a* reports statistics for the full sample, and set *b* reports statistics for the 1991–1993 housing sample.[2]

Recall that our pay data do not contain a random sample of enlisted personnel but data from 1985 through 1993 for individuals from different entering cohorts. Note that the full sample and the housing sample differ in predictable ways, given that the housing sample is restricted to fiscal years 1991–1993: The two samples differ across variables that we expect to change as cohorts age. For example, one result of the housing sample being drawn from later fiscal years is that members are more likely to have dependents in the housing sample than in the full sample. In the full sample, 52 percent of respondents in the Army have no dependents, whereas, in the housing sample, 43 percent have no dependents.

Another characteristic that differs across the samples is years of service. The average years of service for the full sample is between four and five for the Army, Navy, and Marines; it is over six for the Air Force. For the housing sample, the average years of service is higher: nearly six for the Marines, about six and two-thirds for the Army and Navy, and over eight for the Air Force.

---

[1]Appendix E defines all variables.

[2]All statistics were computed using sample observations that were weighted to account for the fact that we selected a nonrandom sample of service members. See the data description in Section 3 for more details.

Table 5.1a

Mean and Standard Deviation of Sample Individual Characteristics for Full Sample, by Service

Characteristic	Army	Navy	Marines	Air Force
Years of Service	4.94	4.87	4.55	6.15
	(7.49)	(7.50)	(7.45)	(8.48)
Age at Entry	19.50	19.51	18.92	19.39
	(5.3)	(5.54)	(4.00)	(4.31)
White Female	0.06	0.08	0.03	0.11
	(0.51)	(0.61)	(0.43)	(0.72)
White Male	0.58	0.66	0.69	0.69
	(1.10)	(1.08)	(1.10)	(1.05)
Black Female	0.05	0.03	0.02	0.04
	(0.50)	(0.39)	(0.30)	(0.42)
Black Male	0.26	0.17	0.19	0.13
	(0.97)	(0.85)	(0.94)	(0.76)
Other Female	0.01	0.01	0.00	0.01
	(0.16)	(0.16)	(0.15)	(0.17)
Other Male	0.04	0.06	0.06	0.03
	(0.46)	(0.52)	(0.57)	(0.40)
0 Dependents	0.52	0.58	0.57	0.45
	(1.11)	(1.12)	(1.18)	(1.13)
1 Dependent	0.18	0.18	0.20	0.22
	(0.86)	(0.87)	(0.96)	(0.94)
2 Dependents	0.15	0.13	0.12	0.16
	(0.78)	(0.76)	(0.78)	(0.83)
3 Dependents	0.10	0.08	0.07	0.12
	(0.68)	(0.61)	(0.62)	(0.73)
4+ Dependents	0.04	0.03	0.03	0.05
	(0.45)	(0.41)	(0.44)	(0.50)
Infantry, Combat	0.29	0.08	0.28	0.08
	(1.00)	(0.63)	(1.07)	(0.60)
Electronic Equipment Repairer	0.04	0.11	0.06	0.11
	(0.45)	(0.71)	(0.60)	(0.74)
Communications/ Intelligence Specialist	0.14	0.12	0.07	0.06
	(0.76)	(0.73)	(0.65)	(0.54)
Health Care Specialist	0.06	0.07	—	0.06
	(0.53)	(0.58)	—	(0.56)
Other Technical and Allied Specialists	0.02	0.01	0.02	0.04
	(0.32)	(0.21)	(0.35)	(0.43)
Functional Support and Administration	0.14	0.10	0.13	0.21
	(0.76)	(0.69)	(0.83)	(0.92)
Electrical/Mechanical Equipment Repairer	0.17	0.24	0.17	0.26
	(0.82)	(0.97)	(0.90)	(0.99)
Craftsman	0.02	0.07	0.02	0.06
	(0.33)	(0.58)	(0.38)	(0.53)
Service and Supply Handler	0.12	0.06	0.14	0.10
	(0.72)	(0.53)	(0.85)	(0.68)
Non Occupational	0.01	0.14	0.06	0.03
	(0.21)	(0.78)	(0.58)	(0.37)
AFQT Percentile	51.19	52.09	56.63	60.02
	(46.97)	(46.95)	(47.22)	(44.73)

Table 5.1a—continued

Characteristic	Army	Navy	Marines	Air Force
AFQT Unknown	0.00	0.01	0.00	0.01
	(0.14)	(0.27)	(0.13)	(0.22)
1978 Cohort	0.08	0.06	0.06	0.12
	(0.59)	(0.53)	(0.56)	(0.74)
1980 Cohort	0.09	0.09	0.07	0.13
	(0.65)	(0.65)	(0.61)	(0.77)
1982 Cohort	0.11	0.11	0.10	0.15
	(0.70)	(0.71)	(0.72)	(0.82)
1984 Cohort	0.20	0.17	0.22	0.19
	(0.89)	(0.86)	(0.98)	(0.90)
1986 Cohort	0.21	0.22	0.19	0.20
	(0.90)	(0.94)	(0.94)	(0.92)
1988 Cohort	0.17	0.21	0.20	0.11
	(0.84)	(0.93)	(0.97)	(0.72)
1990 Cohort	0.13	0.14	0.16	0.08
	(0.74)	(0.78)	(0.88)	(0.63)

NOTE: Standard deviation is in parentheses.

Finally, individuals in the housing sample are more likely to be from later cohorts than are individuals in the full sample. Other characteristics, such as the race and gender distribution, the occupational distribution, and the AFQT distribution, show only slight variations across the two samples.

Contribution of the Components to the Compensation Measures

Since Basic Pay contains only one component, we analyzed the contribution of the components to the levels and variance of RMC, Cash Compensation, and Total Compensation measures. Because YOS affects the distribution of components,[3] we report findings by YOS, but only for calendar year 1991.[4]

Basic Pay and Cash Compensation

Table 5.4 reports the fraction of enlisted personnel receiving the various components. First, as noted earlier, not all personnel receive every benefit, which

---

[3]We stopped at 14 years of service because this was the highest year of service for which we had substantial numbers of observations, given that we had no data after 1993 and our earliest cohort enlisted in 1978.

[4]The results for other calendar years were substantively similar. We chose to present results for 1991, since that fiscal year covered the broadest band of years of service and cohorts in our data set. Even though 1991 was in the drawdown period, the results from that year are representative of the results for other years.

Table 5.1b

**Mean and Standard Deviation of Sample Individual Characteristics for
the Housing Sample, by Service**

Characteristic	Army	Navy	Marines	Air Force
Years of Service	6.66	6.60	5.94	8.35
	(8.63)	(8.44)	(8.74)	(9.49)
Age at Entry	19.53	19.55	18.94	19.38
	(5.34)	(5.65)	(3.97)	(4.30)
White Female	0.05	0.08	0.03	0.11
	(0.47)	(0.59)	(0.40)	(0.71)
White Male	0.55	0.64	0.69	0.69
	(1.09)	(1.08)	(1.10)	(1.05)
Black Female	0.06	0.03	0.02	0.04
	(0.52)	(0.41)	(0.30)	(0.43)
Black Male	0.28	0.18	0.19	0.13
	(0.98)	(0.87)	(0.94)	(0.76)
Other Female	0.01	0.01	0.06	0.01
	(0.17)	(0.17)	(0.16)	(0.18)
Other Male	0.05	0.06	0.07	0.03
	(0.48)	(0.54)	(0.60)	(0.40)
0 Dependents	0.43	0.46	0.46	0.35
	(1.08)	(1.12)	(1.19)	(1.08)
1 Dependent	0.19	0.19	0.22	0.22
	(0.87)	(0.89)	(0.99)	(0.94)
2 Dependents	0.17	0.16	0.16	0.19
	(0.82)	(0.83)	(0.86)	(0.89)
3 Dependents	0.14	0.12	0.11	0.17
	(0.76)	(0.74)	(0.75)	(0.85)
4+ Dependents	0.07	0.06	0.05	0.08
	(0.56)	(0.54)	(0.53)	(0.63)
Infantry, Combat	0.31	0.11	0.30	0.07
	(1.02)	(0.69)	(1.09)	(0.59)
Electronic Equipment Repairer	0.04	0.15	0.08	0.12
	(0.45)	(0.79)	(0.64)	(0.73)
Communications/Intelligence Specialist	0.13	0.13	0.08	0.07
	(0.74)	(0.75)	(0.65)	(0.57)
Health Care Specialist	0.07	0.08	—	0.08
	(0.55)	(0.61)	—	(0.61)
Other Technical and Allied Specialists	0.02	0.01	0.03	0.04
	(0.31)	(0.21)	(0.38)	(0.45)
Functional Support and Administration	0.14	0.11	0.15	0.22
	(0.75)	(0.70)	(0.85)	(0.94)
Electrical/Mechanical Equipment Repairer	0.16	0.26	0.18	0.25
	(0.80)	(0.99)	(0.91)	(0.99)
Craftsman	0.02	0.08	0.03	0.06
	(0.31)	(0.60)	(0.41)	(0.53)
Service and Supply Handler	0.11	0.06	0.14	0.09
	(0.68)	(0.54)	(0.84)	(0.67)
Non Occupational	0.00	0.02	0.01	0.0
	(0.14)	(0.30)	(0.24)	(0.15)
AFQT Percentile	51.79	52.79	57.79	60.99
	(45.55)	(46.89)	(46.05)	(43.88)

Table 5.1b—continued

Characteristic	Army	Navy	Marines	Air Force
AFQT Unknown	0.01	0.02	0.00	0.01
	(0.16)	(0.30)	(0.12)	(0.22)
1978 Cohort	0.07	0.05	0.05	0.12
	(0.56)	(0.51)	(0.52)	(0.74)
1980 Cohort	0.08	0.08	0.06	0.12
	(0.59)	(0.60)	(0.55)	(0.73)
1982 Cohort	0.08	0.08	0.06	0.11
	(0.61)	(0.60)	(0.58)	(0.72)
1984 Cohort	0.11	0.10	0.09	0.14
	(0.67)	(0.66)	(0.69)	(0.78)
1986 Cohort	0.12	0.14	0.09	0.15
	(0.71)	(0.77)	(0.69)	(0.81)
1988 Cohort	0.20	0.24	0.23	0.15
	(0.87)	(0.96)	(1.01)	(0.80)
1990 Cohort	0.35	0.32	0.41	0.21
	(1.04)	(1.04)	(1.17)	(0.93)

NOTE: Standard deviation is in parentheses.

Table 5.2a

**Mean and Standard Deviation of Monthly Compensation Components
(in dollars) for Full Sample, by Service**

Component	Army	Navy	Marines	Air Force
Basic Pay	1,140.18	1,120.64	1,092.03	1,184.03
	(519.60)	(543.45)	(522.75)	(521.32)
BAQ	137.62	153.04	137.74	157.96
	(391.15)	(386.67)	(418.16)	(381.71)
BAS	96.22	86.29	76.96	156.05
	(219.74)	(216.81)	(230.21)	(194.80)
Special Pays	67.29	104.82	58.97	75.12
	(211.46)	(339.70)	(245.48)	(287.34)
Enlistment/	14.42	11.47	237.11	49.38
Reenlistment Bonus	(241.90)	(227.19)	(2,172.68)	(412.55)
Medical Benefits	255.70	242.82	241.03	274.54
	(224.90)	(224.94)	(235.07)	(222.85)
Retirement Benefit	1,543.87	1,540.30	1,304.45	1,970.21
	(2,079.63)	(2,183.74)	(2,872.36)	(2,662.33)
Tax Advantage	56.06	66.06	64.87	81.75
	(149.56)	(166.22)	(225.38)	(188.00)

NOTE: Standard deviation is in parentheses.

accounts for some of the variation in compensation. While all service members receive Basic Pay, receipt of the other components varies considerably. Across all years of service, the percentage of individuals receiving some Special Pays ranges from 57 to 61 percent. In contrast to Special Pays, the percentage of personnel earning the other components differs more by year of service. For the

Table 5.2b

**Mean and Standard Deviation of Monthly Compensation Components (in dollars)
for Housing Sample, by Service**

Component	Army	Navy	Marines	Air Force
Basic Pay	1,229.63	1,218.31	1,159.15	1,283.11
	(555.68)	(562.39)	(581.71)	(564.31)
BAQ	166.54	193.29	149.96	182.53
	(444.56)	(402.18)	(473.09)	(412.84)
BAS	129.00	101.66	101.23	167.38
	(219.00)	(201.66)	(234.29)	(168.51)
Special Pays	92.11	130.52	65.71	71.76
	(281.23)	(382.75)	(300.31)	(299.45)
Enlistment/Reenlistment	56.79	17.49	122.47	132.57
Bonus	(462.17)	(273.20)	(824.03)	(650.04)
Housing	257.07	200.84	283.14	236.34
	(571.37)	(564.54)	(646.90)	(620.01)
Medical Benefits	276.36	270.40	264.15	298.91
	(222.27)	(227.51)	(241.11)	(212.52)
Retirement Benefit	1,791.01	1,636.83	1,479.53	1,969.43
	(1,984.01)	(1,979.00)	(3,009.95)	(2,022.36)
Tax Advantage	69.61	79.30	66.71	85.78
	(182.77)	(173.24)	(222.25)	(189.11)

NOTE: Standard deviation is in parentheses.

Table 5.3a

**Mean and Standard Deviation of Monthly Compensation Measures (in dollars)
for Full Sample, by Service**

Compensation Measure	Army	Navy	Marines	Air Force
Basic Pay	1,140.18	1,120.64	1,092.03	1,184.03
	(519.60)	(543.44)	(522.75)	(521.32)
RMC	1,430.09	1,426.03	1,371.61	1,579.80
	(998.51)	(1,036.08)	(1,144.73)	(973.98)
Cash Compensation	1,455.60	1,476.11	1,602.68	1,622.54
(Without Housing)	(1,025.21)	(1,093.94)	(2,595.76)	(1,123.48)

NOTE: Standard deviation is in parentheses.

Enlistment/Reenlistment Bonus and BAS, receipt is more likely as year of service rises. The incidence of both BAQ payments and In-Kind Housing falls with years of service. This may seem counterintuitive, but recall that members without dependents living in base housing receive partial BAQ. As years of service increase, fewer service members are without dependents, reducing the fraction of individuals eligible to receive both BAQ and base housing.[5]

---

[5]An alternative to treating BAQ and On-Base Housing as two separate benefits would be to treat them as one housing benefit. In this case, 100 percent of service members receives the combined Housing Benefit. The mean benefit levels would be the sum of the means of BAQ and On-Base

Table 5.3b

**Mean and Standard Deviation of Monthly Compensation Measures (in dollars)
for Housing Sample, by Service**

Compensation Measure	Army	Navy	Marines	Air Force
Basic Pay	1,229.63	1,218.31	1,159.15	1,283.11
	(555.68)	(562.39)	(581.71)	(564.31)
RMC	1,589.54	1,592.57	1,477.04	1,718.79
	(1,055.42)	(1,019.45)	(1,172.36)	(992.51)
Cash Compensation	1,668.78	1,661.22	1,598.55	1,837.30
(Without Housing)	(1,179.11)	(1,072.86)	(1,572.66)	(1,308.14)
Cash Compensation	1,925.86	1,862.07	1,881.69	2,073.64
(Including Housing)	(1,151.41)	(1,028.15)	(1,519.47)	(1,285.23)
Total Compensation	4,062.84	3,848.61	3,692.08	4,427.74
	(3,039.32)	(2,876.21)	(4,259.75)	(2,907.93)

NOTE: Standard deviation is in parentheses.

Table 5.4

**Percentage of Enlisted Personnel Receiving Selected Components,
by Year of Service in 1991**

Component	Year of Service							
	2	4	6	8	10	12	14	All
Basic Pay	100.0	100.0	100.0	100.0	100.0	100.0	100.0	100.0
BAQ	96.5	90.9	76.0	70.4	68.5	64.4	64.2	82.5
BAS	33.7	51.1	75.1	83.1	86.8	88.5	91.7	61.7
Housing	77.2	59.6	46.2	41.6	37.8	40.4	38.5	55.8
Special Pays	60.1	61.2	59.5	56.5	59.3	58.7	57.9	58.9
Enlistment/ Reenlistment								
Bonus	0.6	1.1	8.1	22.5	27.2	25.5	32.2	10.0

Next, we report the mean monthly benefit for those receiving the components, by year of service in 1991. As shown in Table 5.5, the value of each benefit rises with year of service. Basic Pay increases at a decreasing rate across years of service, with the exception of the change from year 12 to year 14. The other components show similar slowdowns in growth across years of service, except BAS, which is relatively constant across YOS. BAQ exhibits the largest percentage increases in the early years of service, most likely reflecting the shift of personnel from having no dependents and receiving partial BAQ in the early years to moving off base and receiving BAQ at the full rate in later years.

In Table 5.6, we report the mean monthly benefit for the entire sample of individuals in 1991, irrespective of receipt of the components. The mean value in

Housing. Treating them as one benefit has different implications for our description of the variance of compensation, which we discuss later in this section.

Table 5.5

Mean Monthly Benefit (in dollars) for Those Receiving Selected Components, by Year of Service in 1991

Component	Year of Service							
	2	4	6	8	10	12	14	All
Basic Pay	896	1,061	1,189	1,301	1,396	1,492	1,617	1,153
BAQ	75	144	238	297	344	361	389	187
BAS	165	173	181	186	186	186	189	180
Housing	348	391	482	550	591	625	666	435
Special Pays	85	125	163	166	192	202	201	136
Enlistment/ Reenlistment Bonus	231	327	410	618	608	598	658	580

NOTE: All values are in 1992 dollars.

Table 5.6

Mean Monthly Benefit, Including Individuals Not Receiving Benefit (in dollars), by Year of Service in 1991

Component	Year of Service							
	2	4	6	8	10	12	14	All
Basic Pay	896	1,061	1,189	1,301	1,396	1,492	1,617	1,153
BAQ	73	131	181	209	235	232	250	155
BAS	56	88	136	154	161	165	173	111
Housing	269	233	223	229	224	252	257	243
Special Pays	51	76	97	94	114	119	116	80
Enlistment/ Reenlistment Bonus	1	4	33	139	166	153	212	58

NOTE: All values are in 1992 dollars.

this table combines both incidence and value of the components. Since the incidence of receipt and the mean monthly benefit for those receiving the components rose across years of service for BAS, Special Pays, and the Enlistment/Reenlistment Bonus, the mean monthly benefit irrespective of receipt in Table 5.6 shows more-dramatic percentage increases in the early years, with the rate of growth flattening out in the later years. The receipt of both BAQ and Housing drops with year of service; therefore, even though the mean monthly benefit for those receiving these components rises with year of service, the net mean monthly benefit for BAQ and Housing in Table 5.6 is relatively flat after a few years of service. Since 100 percent of individuals receive Basic Pay, Table 5.6 reports the same mean monthly Basic Pay as Table 5.5.

Table 5.7 reports the mean percentage of RMC that is attributable to each component, by year of service and across all years of service in 1991. For each year of service and across all years of service, Basic Pay accounts for the largest

Table 5.7

Mean Percentage of RMC, by Year of Service in 1991

	Year of Service							
Component	2	4	6	8	10	12	14	All
Basic Pay	88.0	82.6	77.3	75.9	75.2	76.3	76.1	81.0
BAQ	5.4	8.2	10.1	10.8	11.3	10.6	10.6	8.6
BAS	4.4	5.9	8.3	8.7	8.5	8.3	8.1	6.7
Tax Advantage	2.2	3.3	4.2	4.6	5.0	4.9	5.2	3.6

fraction of RMC. This fraction declines with year of service, ranging from a high of 88 percent for those with two years of service to 75 to 76 percent of RMC for those with eight or more years of service. BAQ adds the next-highest amount, on average, to mean RMC—except that the fraction grows rather than declining with year of service. By 12 years of service, BAQ makes up over 10 percent of RMC. The percentage of RMC due to BAS also rises across years of service initially, but levels out at around 8 percent of RMC after six years of service.

Table 5.8 reports the percentage of total variance in RMC, within different years of service and across all years of service, that is due to the variance in each component. The variance within a year indicates the sources of pay differences between service members of the same year of service. The variance across years indicates the components of pay that are sources of pay differences across all years of service in our data. While Basic Pay accounts for the bulk of the mean of RMC, it accounts for a small fraction of the variance in RMC. Within a year of service, Basic Pay never contributes more than 13 percent of the total variance in RMC. Across years of service, Basic Pay is 28 percent of the variance.

BAQ represents a larger fraction of the variance within a year of service than Basic Pay, but it accounts for a smaller fraction of the variance in pay across all years of service. In other words, service members of the same year of service

Table 5.8

Percentage of Variance in RMC, by Year of Service in 1991

	Year of Service							
Component	2	4	6	8	10	12	14	All
Basic Pay	2.9	4.7	4.8	8.8	9.2	12.1	13.4	28.0
BAQ	31.2	33.7	42.5	43.1	41.3	41.9	39.6	15.3
BAS	15.2	12.7	10.3	7.5	5.8	4.7	3.1	4.5
Tax Advantage	3.1	3.7	4.9	6.5	9.2	9.0	11.0	2.5
Covariance Terms	47.5	45.2	37.5	34.1	34.5	32.4	32.9	49.6

who earn different amounts are likely to earn different amounts because of differences in the BAQ they receive, with little difference in their Basic Pay. However, across years of service, service members with different pay are likely to receive different amounts of Basic Pay and, to a lesser extent, different amounts of BAQ.[6] The covariance terms total just under half of the variance in RMC across all years of service and range from 32 to 48 percent of within-year variance. In other words, a large part of the total variance occurs because the various components rise or fall in tandem with each other. Recall that the covariance term is that portion of the variance explained by the covariance in the components with each other. This term could also be viewed as the *residual* of the variance: whatever is not explained by the variances of the individual components.

We indicate the contribution of the components of compensation to the level of the Cash Compensation measure in Table 5.9. The top value in each cell indicates the mean percentage the component contributes to Cash Compensation without the value of In-Kind Housing; the bottom value indicates the mean percentage that the component contributes when the In-Kind Housing value is included. Across all years of service, service members derive more than 70 percent of Cash Compensation from Basic Pay when the Housing value is excluded and 62 percent or more of Cash Compensation when the Housing value is included. As year of service increases, Basic Pay constitutes a smaller portion of Cash Compensation. The next-largest contribution to Cash Compensation comes from BAQ, which accounts for an average of only about 8.5 percent. BAS ranges from about 4 to 8 percent of Cash Compensation on average. The other components were generally less than 5 percent of Cash Compensation.

As Table 5.10 bears out again, although Basic Pay is the largest contributor to Cash Compensation, much of the variation in Cash Compensation comes from sources other than Basic Pay. The figures indicate that Basic Pay contributes relatively little—between 4 and 10 percent—to the total variance in pay within a year of service. Across all years of service, Basic Pay accounts for nearly one-fifth of the variance in pay across members, illustrating that Basic Pay explains relatively more of the differences in pay for members with different years of service.

---

[6]In the preceding footnote, we mentioned the possibility of treating BAQ and On-Base Housing as one housing benefit rather than as two separate benefits. The variance of this combined benefit is a smaller percentage than the variance of either BAQ or On-Base Housing for every year of service and across years of service. Since service members receive no BAQ or small BAQ when they receive On-Base Housing and vice versa, their covariance is negative and large. Treating the two as one benefit, therefore, substantially influences the covariance term: It becomes positive rather than negative within all years of service and stays positive across years of service. These results are available from the authors.

Table 5.9

Mean Percentage of Cash Compensation, by Year of Service in 1991:
Without Value of In-Kind Housing
(Including Value of In-Kind Housing)

	Year of Service							
Component	2	4	6	8	10	12	14	All
Basic Pay	85.4	80.3	74.6	71.1	69.9	71.2	70.5	77.8
	(67.2)	(67.2)	(65.0)	(62.8)	(62.4)	(63.3)	(63.2)	(65.5)
BAQ	5.5	8.3	10.1	10.4	10.8	10.2	10.1	8.6
	(5.3)	(8.3)	(10.1)	(10.4)	(10.8)	(10.2)	(10.1)	(8.5)
BAS	4.4	6.0	8.2	8.3	8.0	7.8	7.5	6.6
	(3.8)	(5.3)	(7.3)	(7.3)	(7.1)	(6.9)	(6.7)	(5.8)
Special Pays	4.5	5.2	5.5	4.8	5.3	5.2	4.7	4.8
	(3.6)	(4.5)	(4.8)	(4.2)	(4.7)	(4.6)	(4.2)	(4.1)
Enlistment/ Reenlistment Bonus	0.1	0.2	1.5	5.5	6.1	5.5	7.1	2.2
	(0.1)	(0.2)	(1.4)	(5.0)	(5.6)	(5.0)	(6.5)	(2.0)
Housing	0.0	0.0	0.0	0.0	0.0	0.0	0.0	0.0
	(19.9)	(14.5)	(11.5)	(10.2)	(9.3)	(9.9)	(9.2)	(14.0)

Table 5.10

Percentage of Variance in Cash Compensation, by Year of Service in 1991:
Without Value of In-Kind Housing
(Including Value of In-Kind Housing)

	Year of Service							
Component	2	4	6	8	10	12	14	All
Basic Pay	3.6	5.5	4.1	4.0	4.1	6.0	5.9	20.2
	(6.3)	(9.6)	(4.7)	(4.1)	(4.3)	(6.0)	(5.7)	(21.8)
BAQ	38.9	39.0	36.3	19.8	18.3	20.7	17.4	11.1
	(67.9)	(68.4)	(42.2)	(20.2)	(19.1)	(20.5)	(16.9)	(11.9)
BAS	19.0	14.7	8.8	3.5	2.6	2.3	1.4	3.2
	(33.1)	(25.8)	(10.3)	(3.5)	(2.7)	(2.2)	(1.3)	(3.5)
Special Pays	13.6	18.6	23.7	10.6	12.0	14.6	11.1	5.3
	(23.8)	(32.6)	(27.5)	(10.8)	(12.5)	(14.4)	(10.9)	(5.7)
Enlistment/ Reenlistment Bonus	1.2	2.8	23.7	66.3	65.8	58.6	67.3	17.7
	(2.1)	(4.9)	(27.5)	(67.8)	(68.6)	(58.1)	(65.2)	(19.0)
Housing	0.0	0.0	0.0	0.0	0.0	0.0	0.0	0.0
	(126.4)	(122.7)	(94.8)	(52.4)	(47.3)	(55.0)	(46.6)	(23.6)
Covariance Terms	23.7	29.8	3.3	−4.2	−2.8	−2.2	−3.2	42.5
	(−159.6)	(−164.0)	(−107.1)	(−58.9)	(−54.4)	(−56.3)	(−46.7)	(14.4)

At higher years of service, the Enlistment/Reenlistment Bonus, at over 50 percent, accounts for the largest share of the variance within a year of service compared with less than 7 percent of the level of Cash Compensation. Across all years of service, the variance in the Enlistment/Reenlistment Bonus represents nearly one-fifth of the variance in Cash Compensation. Housing and BAQ make up the next-largest contributions to the variance within a year and across years of service. The covariance term accounts for over two-fifths of the variance across all years.

We present similar statistics in Tables 5.11 and 5.12, but we replace Cash Compensation with our Total Compensation measure. With the added benefits, Basic Pay now accounts for only one-quarter to one-half of service members' pay. In early years of service, Basic Pay is the largest contributor to Total Compensation, but in later years the accruing Retirement Benefit accounts for the largest share of Total Compensation. Medical Benefits make up about 7 to 10 percent of Total Compensation, about in line with what might be observed in the private sector. Except in the early years of service, when In-Kind Housing is an important contributor, the remaining components each account for less than 5 percent of Total Compensation. Also note that, in the early years of service, cash components account for over one-half the level of Total Compensation; however, in the later years of service, noncash components, such as the Retirement Benefit, Medical Benefits, and In-Kind Housing, make up the bulk of Total Compensation.

Table 5.11

Mean Percentage of Total Compensation, by Year of Service in 1991

Component	Year of Service							
	2	4	6	8	10	12	14	All
Basic Pay	45.3	37.7	27.6	27.1	26.8	29.1	30.2	34.8
BAQ	3.3	4.4	4.2	4.4	4.5	4.6	4.7	4.1
Housing	13.8	8.5	5.1	4.6	4.2	4.8	4.6	8.1
BAS	2.5	3.0	3.1	3.2	3.0	3.2	3.2	2.9
Special Pays	2.5	2.5	2.1	1.9	2.1	2.2	2.1	2.2
Enlistment/ Reenlistment Bonus	0.1	0.1	0.7	2.5	2.7	2.5	3.5	1.0
Tax Advantage	1.3	1.7	1.7	1.8	1.9	2.0	2.2	1.7
Medical Benefits	9.7	8.1	6.6	6.5	6.4	6.7	6.6	7.8
Retirement Benefit	21.5	33.9	49.0	48.0	48.1	44.9	42.7	37.4

Table 5.12

Percentage of Variance in Total Compensation, by Year of Service in 1991

| | Year of Service | | | | | | | |
Component	2	4	6	8	10	12	14	All
Basic Pay	2.2	1.3	1.1	1.3	2.3	3.1	4.0	2.7
BAQ	23.5	9.6	10.0	6.2	10.1	10.8	11.8	1.5
Housing	43.7	17.2	22.5	16.1	25.1	29.0	32.7	3.0
BAS	11.4	3.6	2.4	1.1	1.4	1.2	0.9	0.4
Special Pays	8.2	4.6	6.5	3.3	6.6	7.6	7.6	0.7
Enlistment/ Reenlistment Bonus	0.7	0.7	6.5	20.8	36.3	30.6	45.7	2.4
Tax Advantage	2.3	1.1	1.2	0.9	2.2	2.3	3.3	0.2
Medical Benefits	9.4	3.5	2.9	1.4	1.5	1.4	1.1	0.5
Retirement Benefit	26.6	69.2	61.2	55.7	24.7	24.9	29.0	45.8
Covariance Terms	−28.1	−10.7	−14.4	−6.8	−10.4	−11.1	−36.2	42.8

Table 5.12 demonstrates that, as with Cash Compensation, the components explaining the level of Total Compensation do not explain its variance. While Basic Pay accounts for roughly one-third of the level of Total Compensation both within a given year of service and across years of service, it contributes 4 percent or less to the variance. The Retirement Benefit is the component that accounts for the largest fraction of Total Compensation variance across years of service and for a large part of the variance within a year of service.[7] The Enlistment/ Reenlistment Bonus variance is again a large percentage of the total variance within later years of service but makes up little of the variance across years of service. Variance in the Housing benefit also accounts for a large share of Total Compensation variance within years—but not across years. Especially in the early years of service, BAQ adds a large amount to the within-year variance, but adds less than 2 percent to the across-years variance. Starting in the fourth year of service, all the other components' individual variances are less than 10 percent of Total Compensation variance whether within a year of service or across years.

Summary

Together, Tables 5.4 through 5.12 indicate that the components that make up the bulk of enlisted compensation account for a smaller portion of the variance in

---

[7]We observe a U-shaped pattern for the percentage of the levels and variance in Total Compensation for the Retirement Benefit because of the way the probability of retirement changes over years of service. Recall that the Retirement Benefit is the addition to the expected retirement value in each year. The probability of retirement does not *vary much across individuals* in later years, so the added value of staying additional years has a small variance.

enlisted compensation than they contribute to the levels. While the largest contributor to the levels of Cash Compensation is Basic Pay, the largest contributor to variance in Cash Compensation is the Enlistment/Reenlistment Bonus. For Total Compensation, Basic Pay and the Retirement Benefit account for the majority of the levels, whereas the Retirement Benefit and the Enlistment/Reenlistment Bonus contribute the most to the variance.

This exercise indicates that enlistees' take-home pay is largely determined by the amount of Basic Pay they receive, but that if some service members are receiving more take-home pay than others, this difference is likely to be explained by differences in Enlistment/Reenlistment Bonuses. For Total Compensation, with the exception of Enlistment/Reenlistment Bonuses, most differences in pay are explained by noncash components.

The implication of these comparisons is that while Basic Pay is the compensation tool used to establish the average compensation of individuals, it is not the compensation component used to direct more or less compensation to particular individuals. Variation from the mean of compensation is due to bonus payments, accrual of future retirement benefits, and noncash benefits.

Contribution of Characteristics to the Variation in Components

Next, we investigated the contribution of service-member characteristics to the differentiation in components across members. As discussed earlier, to parcel out the contribution to components' levels and variances to particular characteristics, we ran regressions of each component on the set of service-member characteristics that are likely to explain the levels and variances of components.[8]

It was not evident a priori whether these variables should all influence earnings in a strictly linear fashion. That is, there were no clear theoretical reasons for expecting a characteristic, such as which branch of service the individual is in, to contribute a constant amount to each individual's earnings versus the alternative that that service would raise the contribution to earnings of one or more characteristics such as years of service.

Because of the possibility that the characteristics influenced components and the compensation measures in a nonlinear fashion, we examined the possibility of running the equations separately by gender, race, cohort, and service. We also

---

[8]The motivation for our choice of characteristics is documented in Section 4.

tried interacting various characteristics with years of service. Only when equations were run separately by service did we obtain coefficients that resulted in substantively different conclusions.[9] The preferred specification for the components' regressions is one run separately by service.

For each component and compensation measure, we present the estimated regression coefficients, the R^2,[10] the number of observations, and the mean of the dependent variable. We now discuss the important characteristics contributing to each component and compensation measure in turn.

BAQ

From the regression coefficients for BAQ as a function of individual characteristics, in Table 5.13, we can see that, across all four services, the largest differentiation in BAQ is between service members with and without dependents. Having one dependent instead of zero dependents yields an average payment ranging from $131 higher in the Air Force to $224 higher in the Marines. While other characteristics such as being female, having more years of service, and some occupational categories are also associated with higher BAQ, these differences are one-tenth to one-fifth the size of the effect of having one dependent rather than none.

BAS

The differences in BAS by individual characteristics are very similar to those for BAQ: Having dependents is the largest source of variation in BAS; members who have additional years of service, are female, or belong to certain occupational groups also have slightly higher BAS. This similarity to the variation in BAQ by individual characteristics is not surprising, given that both

---

[9]Despite finding that a traditional F-test rejected the hypothesis that all coefficients were equal when equations were run separately by gender, race, cohort, and service, we ran only separate equations by service, for two reasons. First, for gender, race, and cohort, comparing individual coefficients across equations revealed only small coefficient-estimate differences—that is, running regressions separately by these variables did not change conclusions about the sign or magnitude of the effect of a regressor on the outcome variable. However, running equations separately by service resulted in coefficient estimates of different signs or differing markedly in magnitude. Second, our unusually large sample size of over 800,000 observations made it unlikely that an F-test would not reject the hypothesis of different coefficients across equations.

[10]R^2 represents the proportion of variation in the outcome variable "explained" by the variation in the explanatory variables.

Table 5.13

Coefficients of Regression of Monthly BAQ on Individual Characteristics, by Service

Characteristic	Army	Navy	Marines	Air Force
Intercept	−15.28**	−42.91**	37.81**	−7.82*
Years of Service	11.33**	18.50**	16.08**	14.03**
Years of Service Squared	−0.39**	−0.65**	−1.03**	−0.38**
Age at Entry	2.38**	2.22**	2.01**	2.68**
White Female[a]	15.34**	28.35**	15.72**	24.93**
Black Male	−5.04**	1.14	−0.61	−0.92
Black Female	3.61**	9.24**	4.62*	3.80**
Other Male	−5.18**	0.08	−0.09	−2.66*
Other Female	16.85**	16.69**	8.69*	11.82**
1 Dependent[b]	167.76**	207.86**	223.64**	130.75**
2 Dependents	152.03**	185.05**	227.64**	95.87**
3 Dependents	120.59**	137.30**	190.60**	−38.90**
4+ Dependents	83.43**	97.81**	175.74**	6.48**
Electronic Equipment Repairer[c]	−6.05**	−4.08**	7.10**	7.92**
Communications/Intelligence Specialist	−5.21**	−4.97**	0.70	1.34
Health Care Specialist	−4.19**	15.86**	—	5.54**
Other Technical and Allied Specialist	−3.47	−1.44	1.61	2.14
Functional Support and Administration	−11.73**	6.99**	5.08**	3.03**
Electrical/Mechanical Equipment Repairer	−5.57**	6.71**	4.49**	7.35**
Craftsman	−11.84**	3.33**	4.29	−1.39
Service and Supply Handler	−3.55**	−2.28	1.83	−0.06
Non Occupational	−16.71**	−9.95**	5.40**	−21.56**
AFQT Percentile	0.11**	0.24**	0.15**	0.18**
AFQT Unknown	8.16**	15.72**	−4.16	5.56
1980 Cohort[d]	−8.76**	−2.88**	−36.11**	−12.80**
1982 Cohort	−1.94	−3.67**	−64.22**	−9.12**
1984 Cohort	−2.05	−1.64	−88.11**	−15.84**
1986 Cohort	−7.20**	−7.70**	−98.08**	−27.48**
1988 Cohort	−7.62**	−12.38**	−105.00**	−33.97**
1990 Cohort	−8.45**	−11.21**	−101.38**	−38.07**
R-Squared	0.25	0.45	0.55	0.22
Number of Observations	313,022	223,651	98,525	211,771
Mean Monthly BAQ	139.93	159.23	140.88	158.53

[a]Omitted category is White Male.

[b]Omitted category is 0 Dependents.

[c]Omitted category is Infantry, Combat Arms.

[d]Omitted category is 1978 Cohort.

**Significant at 0.01 level; *Significant at 0.05 level.

are allowances whose eligibility criteria are very much alike. Because the results are so similar to those for BAQ, rather than showing the BAS regression results here, we present them in Appendix Table F.1.

Special Pays

In contrast to the patterns for BAQ and BAS, the patterns for Special Pays do not vary substantially by the number of dependents, as Table 5.14 shows. In all four services, Special Pays vary most by occupational category, with the Infantry, Combat Arms group typically receiving more Special Pays than the other occupational categories, although exceptions exist. The differences in Special Pays across occupational categories are smallest in the Army and greatest in the Navy and Air Force. Other than across occupational categories, Special Pays vary little by other individual characteristics, except in the Navy, where years of service adds a substantial amount, on average, to Special Pays and where being female is associated with lower Special Pays.

Enlistment/Reenlistment Bonus

The patterns of variation in Enlistment/Reenlistment Bonus by individual characteristics, reported in Table 5.15, reflect patterns exhibited in tables earlier in this section. For example, Table 5.2a and Table 5.2b reported higher average levels of Enlistment/Reenlistment Bonus for the Marines than for the other services. In Table 5.15, we see this pattern in the substantially higher coefficients for nearly all variables for the Marines than for the Army, Navy, or Air Force. Reflecting the result in Tables 5.5 and 5.6 that mean levels of Enlistment/ Reenlistment Bonus are higher for those with more years of service, we see extremely large coefficients on the Years of Service variable in Table 5.15. Note that relative to the mean level of Enlistment/Reenlistment Bonus in each service, Years of Service contributes the most to Enlistment/Reenlistment Bonus in the Army and Navy.

Another large source of variation in Enlistment/Reenlistment Bonus is occupational differences. In the Army and Marines, nearly every occupational group receives a lower Enlistment/Reenlistment Bonus on average than the Infantry, Combat Arms group, while in the Navy and Air Force some occupational groups receive a lower Enlistment/Reenlistment Bonus and some receive a higher Enlistment/Reenlistment Bonus relative to Infantry, Combat Arms. Consistently across all four services, the Functional Support and Administration occupational group receives a lower Enlistment/Reenlistment Bonus than others.

Table 5.14

Coefficients of Regression of Monthly Special Pays (in dollars) on Individual Characteristics, by Service

Characteristic	Army	Navy	Marines	Air Force
Intercept	−71.90**	33.44**	80.08**	39.13**
Years of Service	16.54**	28.76**	11.22**	18.13**
Years of Service Squared	−0.16**	−1.07**	−0.28	−0.50**
Age at Entry	−0.57**	−1.32**	−3.61**	−3.77**
White Female[a]	−6.67**	−49.54	−0.01**	−1.29
Black Male	−6.33**	−7.07**	1.00	2.46**
Black Female	−10.20**	−52.14**	−4.27**	−2.30
Other Male	−1.29**	3.25**	4.02**	5.91**
Other Female	−8.48**	−42.36**	0.83	2.68
1 Dependent[b]	10.46**	4.70**	1.40	3.54**
2 Dependents	17.52**	7.17**	2.38*	7.35**
3 Dependents	20.76**	11.97**	9.32**	16.63**
4+ Dependents	20.51**	9.80**	0.70	20.02**
Electronic Equipment Repairer[c]	−0.26	−9.85**	−7.88**	−26.46**
Communications/ Intelligence Specialist	2.08**	−3.96**	11.14**	2.62
Health Care Specialist	−11.60**	−60.79**	—	−30.81**
Other Technical and Allied Specialist	−0.13	−45.31**	8.75**	−5.46**
Functional Support and Administration	1.09*	−36.96**	1.31	−23.31**
Electrical/Mechanical Equipment Repairer	2.60**	−25.31**	0.37	−19.84**
Craftsman	−4.57**	−35.12**	5.10*	−27.90**
Service and Supply Handler	2.45**	−18.59**	6.72**	−19.57**
Non Occupational	−18.89**	−46.09**	−13.02**	−36.51**
AFQT Percentile	0.11**	0.12**	0.05**	0.12**
AFQT Unknown	−0.07	25.55**	−8.36	−5.32
1980 Cohort[d]	19.67**	−8.08**	3.75*	9.27**
1982 Cohort	40.76**	15.85**	1.31	21.74**
1984 Cohort	60.14**	14.57**	0.11	35.32**
1986 Cohort	74.91**	24.80**	−0.11	4.05**
1988 Cohort	92.73**	24.64**	−7.64**	54.68**
1990 Cohort	102.26**	22.42**	0.38	56.81**
R-Squared	0.14	0.16	0.08	0.07
Number of Observations	313,022	223,651	98,525	211,771
Mean Monthly Special Pays	66.30	95.98	59.70	73.31

[a]Omitted category is White Male.
[b]Omitted category is 0 Dependents.
[c]Omitted category is Infantry, Combat Arms.
[d]Omitted category is 1978 Cohort.
**Significant at 0.01 level; *Significant at 0.05 level.

Table 5.15

**Coefficients of Regression of Monthly Enlistment/Reenlistment Bonus (in dollars)
on Individual Characteristics, by Service**

Characteristic	Army	Navy	Marines	Air Force
Intercept	−98.08**	−37.15**	−338.43**	−85.27**
Years of Service	7.27**	8.04**	48.88**	−8.10**
Years of Service Squared	0.27**	−0.34**	−4.03**	1.89**
Age at Entry	−0.30**	0.15	3.65**	−1.19**
White Female[a]	−1.52**	−1.24*	−4.47	−5.13**
Black Male	−4.27**	0.82	11.93	3.89**
Black Female	−4.52**	−0.38	22.80	−10.55**
Other Male	−0.70	1.68**	−3.42	1.44
Other Female	−3.12**	−3.05	16.29	−7.10*
1 Dependent[b]	0.26	0.59	24.97**	1.63
2 Dependents	2.25**	0.20	51.05**	5.68**
3 Dependents	1.85**	1.70*	59.04**	6.43**
4+ Dependents	3.74**	2.44*	52.61**	10.50**
Electronic Equipment Repairer[c]	−16.50**	10.82**	232.56**	78.00**
Communications/ Intelligence Specialist	−4.47**	16.82**	−86.21**	118.30**
Health Care Specialist	−13.07**	−1.71	—	−22.48**
Other Technical and Allied Specialist	−3.75**	2.35	−27.76	1.12
Functional Support and Administration	−23.19**	−11.20**	−117.36**	−24.55**
Electrical/Mechanical Equipment Repairer	−17.59**	6.13	−4.96	11.86
Craftsman	−19.20**	7.51**	−60.10**	−26.42**
Service and Supply Handler	−16.91**	4.04**	−115.74**	−25.45**
Non Occupational	−9.25**	8.84**	−28.38*	3.05
AFQT Percentile	0.26**	0.10**	4.02**	0.54**
AFQT Unknown	5.90*	9.86**	52.82	42.10**
1980 Cohort[d]	31.81**	9.15**	−116.81**	37.86**
1982 Cohort	60.59**	8.11**	−109.63**	84.09**
1984 Cohort	79.06**	−2.74	−396.14**	86.37**
1986 Cohort	87.40**	10.97**	−655.22**	79.68**
1988 Cohort	84.15**	7.54**	−705.29**	82.60**
1990 Cohort	94.84**	5.70**	−705.99**	81.65**
R-Squared	0.06	0.02	0.10	0.22
Number of Observations	313,022	223,651	98,525	211,771
Mean Monthly Enlistment/ Reenlistment Bonus	12.38	11.12	224.06	41.95

[a]Omitted category is White Male.

[b]Omitted category is 0 Dependents.

[c]Omitted category is Infantry, Combat Arms.

[d]Omitted category is 1978 Cohort.

**Significant at 0.01 level; *Significant at 0.05 level.

Housing Benefit

Although we used the full sample to estimate the equations for the other components, we estimated the Housing benefit regression using only the housing sample. Table 5.16 shows these results. In contrast to the results for many of the components we have discussed, for the Housing benefit, the presence of dependents lowers rather than raises average values—most likely a result of the fact that individuals with no dependents are more likely to live on base rather than off base and are therefore more likely to get positive rather than zero housing benefits. For similar reasons, we also see large negative coefficients on the dummy variables for females.

Medical Benefits

Regression coefficients for Medical Benefits are shown in Table 5.17. For all four services, the largest source of variation in Medical Benefits is dependent status.[11] All other variables contribute very little variation to Medical Benefits. Note that we are able to explain a large fraction of the total variation in Medical Benefits by the high R^2 for each regression.

Tax Advantage

The results for the Tax Advantage follow the results we obtain for BAQ and BAS: The major sources of differences in Tax Advantage are that females and members with dependents receive higher levels. This result is not surprising, since the Tax Advantage is calculated as the benefit deriving from the nontaxability of allowances and, hence, should reflect the value of allowances. Table 5.18 shows the Tax Advantage coefficients.

Retirement Benefit

The last component we examine is the Retirement Benefit, with results listed in Table 5.19. In all four services, the largest source of differences in the value of

---

[11]An alternative to valuing Medical Benefits the way we did would be to value single members' benefits at one level and those of members with any dependents at another level. This method would be in keeping with firms that offer single health plans at one price, then cover all other additional dependents by a family plan, which has the same price regardless of the number of dependents. Note that valuing Medical Benefits this way would lead to the same substantive conclusion—that the largest source of differences in Medical Benefits was from differences in dependent status—but that there would be none of the additional increase in variation for each additional dependent that we see in Table 5.17.

Table 5.16

**Coefficients of Regression of Monthly Housing Benefit (in dollars) on
Individual Characteristics, by Service**

Characteristic	Army	Navy	Marines	Air Force
Intercept	314.91**	335.83**	360.62**	341.94**
Years of Service	−3.04	−9.23**	−0.52	−13.18**
Years of Service Squared	0.24*	0.45**	0.17	0.44**
Age at Entry	−1.23**	−2.99**	−0.72	−3.58**
White Female[a]	−44.19**	−35.16**	−62.73**	−57.99**
Black Male	−6.61*	−9.88**	−7.70	4.02
Black Female	−17.71**	−1.49	−20.75*	−10.54*
Other Male	7.94*	3.24	8.75	19.44**
Other Female	−36.31**	−19.06*	−27.79	−16.13
1 Dependent[b]	−101.06**	−130.99**	−223.60**	−27.86**
2 Dependents	−62.23**	−81.53**	−172.22**	30.53**
3 Dependents	−5.60	−4.35	−49.24**	116.18**
4+ Dependents	39.53**	48.82**	−35.95**	162.13**
Electronic Equipment Repairer[c]	9.44	10.40*	3.24	−21.46**
Communications/ Intelligence Specialist	5.87	9.27*	12.83*	15.09*
Health Care Specialist	8.96	−16.61**	—	3.84
Other Technical and Allied Specialist	−0.91	21.29	9.30	3.59
Functional Support and Administration	12.81**	−9.46*	8.37	6.16
Electrical/Mechanical Equipment Repairer	−1.92	−18.15**	14.98**	−11.49*
Craftsman	3.31	−3.82	8.71	18.21*
Service and Supply Handler	−4.16	−9.50*	−1.32	10.00
Non Occupational	26.03	−10.18	−19.94	17.29
AFQT Percentile	0.01	−0.08	−0.05	0.00
AFQT Unknown	−20.16	−12.96*	5.10	−1.88
1980 Cohort[d]	12.17*	19.41**	10.15	12.68*
1982 Cohort	4.20	16.67	7.25	6.49
1984 Cohort	5.97	8.00	9.82	10.83
1986 Cohort	11.91	9.59	40.17*	4.54
1988 Cohort	−15.13	6.67	14.85	9.82
1990 Cohort	3.30	7.05	15.22	17.20
R-Squared	0.04	0.06	0.13	0.06
Number of Observations	73,003	63,752	25,798	50,168
Mean Monthly Housing Benefit	252.60	194.91	279.06	221.69

[a]Omitted category is White Male.
[b]Omitted category is 0 Dependents.
[c]Omitted category is Infantry, Combat Arms.
[d]Omitted category is 1978 Cohort.
**Significant at 0.01 level; *Significant at 0.05 level.

Table 5.17

Coefficients of Regression of Monthly Medical Benefits (in dollars) on Individual Characteristics, by Service

Characteristic	Army	Navy	Marines	Air Force
Intercept	147.10**	141.76**	146.47**	143.18**
Years of Service	5.52**	5.46**	2.00**	8.61**
Years of Service Squared	−0.28**	−0.25**	−0.10**	−0.41**
Age at Entry	0.24**	0.32**	0.43**	0.52**
White Female[a]	18.76**	26.61**	16.17**	27.35**
Black Male	1.13**	0.87**	−0.13	−1.11**
Black Female	13.59**	13.64**	11.37**	13.84**
Other Male	0.36**	−0.60*	0.11	−1.78**
Other Female	19.24**	26.55	12.01**	22.72**
1 Dependent[b]	131.87**	133.07**	144.51**	113.41**
2 Dependents	203.75**	205.17**	216.65**	184.82**
3 Dependents	203.46**	203.64**	216.17**	184.20**
4+ Dependents	202.90**	203.90**	216.53**	184.61**
Electronic Equipment Repairer[c]	2.26**	1.36**	2.15**	0.83
Communications/ Intelligence Specialist	1.74**	0.91**	1.70**	0.43
Health Care Specialist	3.14**	3.87**	—	2.95**
Other Technical and Allied Specialist	0.73	−3.95**	0.18	1.52*
Functional Support and Administration	2.34**	2.58**	1.37**	3.12**
Electrical/Mechanical Equipment Repairer	2.11**	0.09	1.56**	0.67**
Craftsman	1.59**	0.41**	2.25**	1.73**
Service and Supply Handler	1.65**	−1.14**	1.10**	1.60**
Non Occupational	−2.73**	−1.37**	−0.43	−7.64**
AFQT Percentile	−0.00	0.00	−0.00	−0.02**
AFQT Unknown	−1.36	−0.43	−2.33	−0.66
1980 Cohort[d]	−0.96**	−0.52	1.00*	−1.78**
1982 Cohort	1.01**	1.40	0.42	0.35
1984 Cohort	0.04	1.27**	−0.05	1.26**
1986 Cohort	−0.89**	0.50	1.20*	1.14**
1988 Cohort	−1.53**	0.30	−0.86	0.10
1990 Cohort	−2.81**	0.16	−0.60	−2.07**
R-Squared	0.89	0.88	0.94	0.79
Number of Observations	313,022	223,651	98,525	211,771
Mean Monthly Medical Benefit	259.03	246.92	243.66	270.70

[a]Omitted category is White Male.

[b]Omitted category is 0 Dependents.

[c]Omitted category is Infantry, Combat Arms.

[d]Omitted category is 1978 Cohort.

**Significant at 0.01 level; *Significant at 0.05 level.

Table 5.18

**Coefficients of OLS Regression of Monthly Tax Advantage (in dollars) on
Individual Characteristics, by Service**

Characteristic	Army	Navy	Marines	Air Force
Intercept	1.77	19.45**	3.08**	24.90**
Years of Service	6.94**	11.45**	11.52**	7.99**
Years of Service Squared	−0.22**	−0.52**	−0.73**	−0.33**
Age at Entry	0.87**	0.65**	0.02	0.43**
White Female[a]	11.03**	20.56**	15.83**	8.33**
Black Male	−2.71**	0.24	−0.21	0.74
Black Female	6.83**	15.39	10.77**	5.29*
Other Male	−0.46	3.98**	3.05**	2.38**
Other Female	12.82**	19.53**	15.46**	2.92*
1 Dependent[b]	50.04**	53.86**	73.42**	38.07**
2 Dependents	44.01**	42.61**	69.80**	23.14**
3 Dependents	31.29**	22.26**	49.91**	1.08
4+ Dependents	19.31**	8.97**	39.47**	−9.97**
Electronic Equipment Repairer[c]	−0.06	−0.51	17.50**	4.17**
Communications/ Intelligence Specialist	0.60	4.14**	1.40	11.74**
Health Care Specialist	1.66**	10.25**	—	−3.20**
Other Technical and Allied Specialist	−0.38	7.03**	6.02**	1.52
Functional Support and Administration	0.20	6.51**	3.44**	−1.20
Electrical/Mechanical Equipment Repairer	−1.14**	1.41*	6.45**	−0.16**
Craftsman	−2.76**	−1.10	2.04	−5.65**
Service and Supply Handler	1.11**	−2.57**	−0.12	−3.01**
Non Occupational	−7.76**	−6.53**	5.18**	−20.74**
AFQT Percentile	0.12**	0.18**	0.27**	0.14**
AFQT Unknown	1.75	12.31**	−0.71	5.32**
1980 Cohort[d]	−8.60**	−11.02**	−40.66**	−15.83**
1982 Cohort	−10.23**	−21.24**	−67.27**	−21.97**
1984 Cohort	−14.71**	−33.39**	−98.31**	−34.61**
1986 Cohort	−20.66**	−40.41**	−111.84**	−45.80**
1988 Cohort	−20.04**	−44.47**	−117.86**	−50.28**
1990 Cohort	−19.17**	−46.29**	−115.06**	−54.89**
R-Squared	.24	.32	.39	.17
Number of Observations	313,022	223,651	98,525	211,771
Mean Monthly Tax Advantage	58.58	71.22	67.07	82.65

[a]Omitted category is White Male.

[b]Omitted category is 0 Dependents.

[c]Omitted category is Infantry, Combat Arms.

[d]Omitted category is 1978 Cohort.

**Significant at 0.01 level; *Significant at 0.05 level.

Table 5.19

**Coefficients of OLS Regression of Monthly Retirement Benefit (in dollars)
on Individual Characteristics, by Service**

Characteristic	Army	Navy	Marines	Air Force
Intercept	959.15**	832.81**	–75.40*	1,460.03**
Years of Service	260.82**	210.83**	403.66**	181.23**
Years of Service Squared	–8.20**	–6.80**	–15.07**	–7.07**
Age at Entry	–21.09**	–16.33**	–8.24**	–16.06**
White Female[a]	–94.58**	–43.33**	–40.81**	–61.60**
Black Male	47.45**	37.03**	102.28**	–60.86**
Black Female	–7.34	–25.62**	41.75**	–49.05**
Other Male	118.65**	96.58**	109.59**	32.84**
Other Female	62.92**	51.44**	129.47**	7.38
1 Dependent[b]	110.42**	39.38**	66.94**	51.39**
2 Dependents	141.82**	96.43**	158.32**	99.67**
3 Dependents	124.51**	95.56**	153.28**	52.94**
4+ Dependents	52.25**	6.17	87.40**	–26.02*
Electronic Equipment Repairer[c]	12.48	111.73**	80.34**	–49.12**
Communications/ Intelligence Specialist	–6.50	173.20**	26.96*	2.61
Health Care Specialist	–6.28	157.02**	—	10.53
Other Technical and Allied Specialist	18.17*	229.30**	93.43**	27.82
Functional Support and Administration	1.93	196.11**	75.11**	66.13**
Electrical/Mechanical Equipment Repairer	15.22**	122.82**	47.54**	19.70
Craftsman	14.60	179.23**	12.60	–18.11
Service and Supply Handler	–13.84**	192.42**	11.26	–25.14*
Non Occupational	559.58**	773.36**	1,211.84**	2,167.49**
AFQT Percentile	0.48**	1.57**	0.78**	0.24
AFQT Unknown	53.93**	–215.78**	–70.48	144.37**
1980 Cohort[d]	–5.55	69.01**	92.81**	69.04**
1982 Cohort	131.56**	109.42**	192.12**	214.52**
1984 Cohort	–248.38**	–294.61**	–187.92**	–204.25**
1986 Cohort	–78.38**	–117.80**	9.60	7.94
1988 Cohort	–121.26**	–54.02**	–46.13*	–46.70**
1990 Cohort	–196.15**	–244.21**	–269.53**	–344.70**
R-Squared	.47	.24	.41	.17
Number of Observations	313,022	223,651	98,525	211,771
Mean Monthly Retirement Benefit	1,573.34	1,551.86	1,335.17	1,955.03

[a]Omitted category is White Male.
[b]Omitted category is 0 Dependents.
[c]Omitted category is Infantry, Combat Arms.
[d]Omitted category is 1978 Cohort.
**Significant at 0.01 level; *Significant at 0.05 level.

this component is from Years of Service, which raises the average level of the Retirement Benefit. This result clearly indicates that when questions regarding remuneration for longevity are being considered, not only current compensation but also promises of future retirement benefits should be factored in. For the Navy, we also see that being in occupational groups other than Infantry, Combat Arms raises average levels of the Retirement Benefit substantially; for the Marines, we observe higher levels of average values for individuals with dependents.

Contribution of Characteristics to Compensation Measures

Finally, we examined the contribution of individual characteristics to the levels of our four compensation measures. We present results from regressions of Basic Pay, RMC, Cash Compensation Without Housing, Cash Compensation Including Housing, and Total Compensation on the set of service-member characteristics.[12]

Basic Pay

The regression results for Basic Pay are reported in Table 5.20. Years of Service exhibits a quadratic effect, with a positive coefficient on the linear term and a negative coefficient on the squared term. This implies that Basic Pay rises with years of service, but at a decreasing rate. In general, women and minorities earn less Basic Pay than white males, but the estimated difference between the earnings of women and minorities and the earnings of men is less than 1 percent of Basic Pay. Individuals with more dependents earn more Basic Pay, but, again, the additional pay due to dependents is small. There are some differences in earnings across occupational groups, but they are also relatively small. AFQT percentile has very little effect on Basic Pay. The successively larger negative coefficients on the cohort dummy variables are consistent with falling real Basic Pay over time, although other factors could have also contributed to this decline.

As indicated by the large R^2 estimates of between .83 and .89, the regressions explain a large portion of the total variance in Basic Pay. However, note that the largest relative contribution to Basic Pay comes not from any of the service-member characteristics but from the Intercepts,[13] which are close to 70 percent of the average values of Basic Pay. This value indicates that the fixed proportion of

---

[12]Please see the subsection "Descriptive Interpretation" in Section 4 for a discussion of the appropriate way to interpret these regression results.

[13]The *intercept* is the expected value of the outcome variable when all the explanatory variables are zero.

Table 5.20

Coefficients of Regression of Monthly Basic Pay (in dollars) on Individual Characteristics, by Service

Characteristic	Army	Navy	Marines	Air Force
Intercept	796.52**	833.06**	831.79**	866.27**
Years of Service	85.57**	76.74**	64.77**	67.65**
Years of Service Squared	−2.13**	−1.62**	−0.80**	−1.32**
Age at Entry	2.84**	0.85**	3.09**	2.62**
White Female[a]	−2.60**	2.47**	6.37**	−3.13**
Black Male	−7.90**	−10.55**	−4.20**	−7.08**
Black Female	−4.50**	−9.06**	−0.80	−1.76*
Other Male	−2.45**	−1.59**	−2.50**	−1.99**
Other Female	−4.61**	−2.26	7.99**	−0.92
1 Dependent[b]	7.72**	5.34**	7.56**	3.31**
2 Dependents	6.34**	8.07**	10.14**	2.92**
3 Dependents	6.54**	11.97**	13.59**	1.66*
4+ Dependents	6.02**	10.27**	13.87**	−2.06*
Electronic Equipment Repairer[c]	−14.60**	10.31**	9.75**	−5.62**
Communications/ Intelligence Specialist	−7.28**	14.45**	4.88**	−4.15**
Health Care Specialist	−17.78**	−17.33	—	−3.25**
Other Technical and Allied Specialist	−4.11**	0.46	6.61**	−4.52**
Functional Support and Administration	−17.34**	6.09**	4.31**	−4.58**
Electrical/Mechanical Equipment Repairer	−12.30**	4.15**	6.32**	−1.47
Craftsman	−13.66**	17.04**	−0.28	−5.32**
Service and Supply Handler	−11.21**	2.19*	−1.58	−2.62**
Non Occupational	−53.44**	−42.77**	−33.52**	−72.14**
AFQT Percentile	0.34**	0.64**	0.46**	0.41**
AFQT Unknown	39.23**	62.25**	55.52**	40.85**
1980 Cohort[d]	−34.53**	−25.56**	−39.09**	−36.52**
1982 Cohort	−39.21**	−40.96**	−61.12**	−42.12**
1984 Cohort	−47.84**	−58.88**	−80.61**	−73.13**
1986 Cohort	−81.34**	−91.69**	−112.43**	−108.64**
1988 Cohort	−101.20**	−126.46**	−144.74**	−124.49**
1990 Cohort	−113.05**	−143.09**	−147.93**	−141.11**
R-Squared	.83	.89	.85	.88
Number of Observations	318,322	227,742	99,557	213,448
Mean Monthly Basic Pay	1,149.4	1,123.9	1,092.8	1,183.2

[a]Omitted category is White Male.
[b]Omitted category is 0 Dependents.
[c]Omitted category is Infantry, Combat Arms.
[d]Omitted category is 1978 Cohort.
**Significant at 0.01 level; *Significant at 0.05 level.

Basic Pay that is not explained by characteristics is very large; in other words, most service members are paid about the same amount, with little of the variation being explained by their characteristics. Years of Service makes the second-largest contribution to the average level of Basic Pay, with an effect between 5 and 7 percent of the average level of Basic Pay for each year of service. Although most of the coefficients on the other characteristics are significant, the magnitude of those coefficients on Basic Pay is very small.

RMC

In Table 5.21 we report the regression results for RMC. Years of Service again exhibits the quadratic pattern we saw for Basic Pay. RMC varies more by race and gender than did Basic Pay: Women in all services earn more RMC on average than their male counterparts, but this premium represents only about 3 to 7 percent of RMC. Whereas Basic Pay varied little by the number of dependents, RMC varies considerably by the number of dependents, with one dependent adding one-fifth to one-quarter to RMC—not surprising, given that RMC adds components to Basic Pay that depend in part on a service member's number of dependents. RMC varies little by occupation and AFQT score. As with Basic Pay, the regression explains a relatively large amount of the total variance in RMC, as shown by the moderately high R^2 values.

Cash Compensation Without Housing

The regression results reported in Table 5.22—with R^2 values ranging from .35 to .74—show that the characteristics explain less of the total variation in Cash Compensation Without Housing than they did for Basic Pay. Nevertheless, Table 5.22 shows that more characteristics make a sizable contribution to Cash Compensation Without Housing than was true for Basic Pay. Here, having dependents makes the largest contribution to earnings. For example, having one dependent leads to earnings from $185 higher in the Air Force to $334 higher in the Marines. Years of Service makes the next-largest contribution to earnings, with an additional year of service adding from $111 in the Air Force to $157 in the Marines, representing from 7 to 10 percent of average Cash Compensation Without Housing.

This time, women earn up to 4 percent more on the whole than white men, and with the exception of the Marines, black men earn a few percent less than white men. The results for Other Male are mixed. Differences in earnings across occupations are larger here than for Basic Pay, with nearly every group except Electronic Equipment Repairer earning slightly less than the reference group,

56

Table 5.21

Coefficients of Regression of Monthly RMC (in dollars) on Individual Characteristics, by Service

Characteristic	Army	Navy	Marines	Air Force
Intercept	783.16**	817.43**	1,918.85**	978.83**
Years of Service	123.52**	127.85**	110.36**	108.71**
Years of Service Squared	–3.62**	–3.87**	–3.29**	–3.01**
Age at Entry	5.12**	3.47**	5.16**	5.18**
White Female[a]	51.38**	99.60**	70.40**	45.80**
Black Male	–21.02**	–14.18**	–12.86**	–11.36**
Black Female	23.07**	61.31**	36.85**	13.61**
Other Male	–9.73**	–0.55	–4.11*	–4.11
Other Female	49.56**	80.94**	58.94**	23.20**
1 Dependent[b]	292.32**	291.23**	381.11**	217.72**
2 Dependents	276.91**	256.11**	394.22**	166.22**
3 Dependents	234.26**	190.45**	339.50**	82.58**
4+ Dependents	184.39**	137.76**	313.57**	31.43**
Electronic Equipment Repairer[c]	–11.36**	8.21**	45.58**	–2.12
Communications/ Intelligence Specialist	–6.49**	20.49**	15.40**	7.76**
Health Care Specialist	–5.24**	52.98**	—	–12.33**
Other Technical and Allied Specialist	–3.06	27.45**	43.12**	–0.45
Functional Support and Administration	–17.58**	36.75**	31.14**	–13.38**
Electrical/Mechanical Equipment Repairer	–15.27**	17.83**	27.68**	–7.74**
Craftsman	–24.97**	11.92**	18.85**	–30.93**
Service and Supply Handler	–0.01	–8.93**	11.07**	–18.33**
Non Occupational	–81.25**	–71.99**	–18.47**	–187.06**
AFQT Percentile	0.72**	1.26**	1.04**	0.76**
AFQT Unknown	57.02**	96.84**	51.45**	52.96**
1980 Cohort[d]	–55.23**	–39.12**	–114.63**	–70.67**
1982 Cohort	–50.81**	–70.34**	–192.99**	–79.28**
1984 Cohort	–63.27**	–98.71**	–270.47**	–131.33**
1986 Cohort	–112.19**	–147.49**	–327.29**	–194.00**
1988 Cohort	–132.86**	–192.93**	–367.59**	–227.43**
1990 Cohort	–138.47**	–214.65**	–364.32**	–260.39**
R-Squared	.68	.75	.78	.67
Number of Observations	313,022	223,651	98,525	211,771
Mean Monthly RMC	1,449.46	1,454.10	1,383.35	1,579.14

[a]Omitted category is White Male.
[b]Omitted category is 0 Dependents.
[c]Omitted category is Infantry, Combat Arms.
[d]Omitted category is 1978 Cohort.
**Significant at 0.01 level; *Significant at 0.05 level.

Table 5.22

Coefficients of Regression of Monthly Cash Compensation Without Housing (in dollars) on Individual Characteristics, by Service

Characteristic	Army	Navy	Marines	Air Force
Intercept	583.07**	799.60**	1,250.54**	868.20**
Years of Service	143.27**	150.04**	157.46**	111.49**
Years of Service Squared	−3.43**	−4.62**	−6.81**	−1.34**
Age at Entry	8.60**	3.64**	11.83**	0.68
White Female[a]	30.32**	28.47**	49.03**	29.90**
Black Male	−30.01**	−21.08**	−0.30	−14.13**
Black Female	−0.27	−5.33**	44.28**	−4.66
Other Male	−12.61**	1.76	−6.25	0.87
Other Female	22.88**	17.52**	62.20*	13.75**
1 Dependent[b]	250.83**	233.68**	333.68**	184.62**
2 Dependents	250.79**	222.27**	376.99**	155.98**
3 Dependents	224.01**	183.69**	358.01**	103.58**
4+ Dependents	187.47**	143.01**	320.44**	71.40**
Electronic Equipment Repairer[c]	−27.56**	10.42**	252.26**	45.08**
Communications/ Intelligence Specialist	−8.35**	31.43**	−59.42**	116.89**
Health Care Specialist	−30.29**	−17.13**	—	−62.10**
Other Technical and Allied Specialist	−5.23	−18.68**	19.20	−16.05**
Functional Support and Administration	−39.15**	−16.14**	−87.03**	−59.97**
Electrical/Mechanical Equipment Repairer	−28.04**	−0.27	17.10**	−14.92**
Craftsman	−44.46**	−13.13**	−37.54*	−79.59**
Service and Supply Handler	−15.38**	−18.38**	−96.78**	−59.45**
Non Occupational	−106.86**	−101.53**	−64.10**	−200.08**
AFQT Percentile	0.98**	1.33**	4.81**	1.28**
AFQT Unknown	60.72**	122.62**	95.94**	87.83**
1980 Cohort[d]	2.06	−26.24**	−190.41**	−8.83**
1982 Cohort	59.39**	−25.95**	−240.19**	48.65**
1984 Cohort	91.75**	−51.89**	−572.19**	25.64**
1986 Cohort	73.30**	−77.25**	−873.85**	−23.82**
1988 Cohort	68.64**	−127.30**	−970.49**	−36.28**
1990 Cohort	84.81**	−152.73**	−961.29**	−63.56**
R-Squared	.69	.74	.35	.67
Number of Observations	318,322	227,742	99,557	213,448
Mean Monthly Cash Compensation Without Housing	1,469.0	1,486.0	1,597.0	1,614.0

[a]Omitted category is White Male.
[b]Omitted category is 0 Dependents.
[c]Omitted category is Infantry, Combat Arms.
[d]Omitted category is 1978 Cohort.
**Significant at 0.01 level; *Significant at 0.05 level.

Infantry, Combat Arms. In contrast to the other services and the results for Basic Pay, Cash Compensation Without Housing in the Army did not steadily decline over successive cohorts.

Cash Compensation Including Housing

The results from adding the value of In-Kind Housing to Cash Compensation are reported in Table 5.23. They are similar to the results for Cash Compensation Without Housing, except for a few aspects. One is that additional dependents do not contribute as much to earnings. The effect of an additional dependent is about half here what it is for Cash Compensation Without Housing. Other differences are that earnings vary slightly more across occupational groups, and real earnings decline less over time for Cash Compensation Including Housing.

Total Compensation

These regressions also explain a large proportion of Total Compensation—between 70 and 84 percent—as shown in Table 5.24. For this measure of compensation, the intercepts are also a large percentage of the mean. Years of Service still makes the biggest contribution to the average level of earnings of the characteristics. An additional year of service adds between $656 per month to Total Compensation in the Navy and an extra $747 per month in the Army. The next-largest explanatory variable is, again, dependents. Individuals with dependents earn up to 16 percent more monthly Total Compensation than their single counterparts. White females earn barely less than white males; blacks earn 2 to 5 percent less than white males. The differences between the earnings of those in the other race categories and white males are insignificant, or less than 2 percent. Differences across occupations remain small—less than 5 percent of the average, with one exception—and the effect of AFQT continues to be negligible.

Summary

These regressions have shown that enlisted compensation does vary with individual characteristics. Although the results for alternative measures of compensation show different patterns, they share some commonalities. First, on the whole, most individuals are receiving about the same average level of compensation, which is illustrated by the large constant terms in each regression. Second, the service-member characteristics usually explaining the biggest difference in earnings are years of service and the presence of dependents. Third, very few other characteristics affect earnings in any amount over 5 percent of the average level.

Table 5.23

Coefficients of Regression of Monthly Cash Compensation Including Housing (in dollars) on Individual Characteristics, by Service

Characteristic	Army	Navy	Marines	Air Force
Intercept	689.44**	1,044.13**	1,381.28**	1,267.96**
Years of Service	150.23**	134.45**	188.18**	97.76**
Years of Service Squared	−2.17**	−2.99**	−7.27**	−0.76**
Age at Entry	−1.34**	−1.67**	3.25*	−5.45**
White Female[a]	−20.70**	−40.00**	−18.23*	−44.80**
Black Male	−45.59**	−36.91**	−27.85**	−23.36**
Black Female	−34.77**	−50.36**	−21.00	−44.06**
Other Male	−5.38	1.59	5.88	14.34**
Other Female	−17.96	−25.13**	7.23	−21.12
1 Dependent[b]	128.24**	75.64**	63.30**	118.82**
2 Dependents	165.99**	108.87**	119.39**	147.22**
3 Dependents	192.29**	154.43**	194.55**	195.16**
4+ Dependents	209.71**	169.22**	173.45**	222.17**
Electronic Equipment Repairer[c]	−74.34**	17.68**	82.92**	112.74**
Communications/ Intelligence Specialist	−17.40**	58.91**	87.87**	305.55**
Health Care Specialist	−67.67**	−48.76**	—	−83.30**
Other Technical and Allied Specialist	−16.30	−10.09	89.26**	−20.57**
Functional Support and Administration	−87.72**	−29.66**	37.02**	−92.99**
Electrical/Mechanical Equipment Repairer	−80.23**	−21.98**	59.40**	−14.75*
Craftsman	−97.17**	1.12	50.45**	−98.10**
Service and Supply Handler	−71.60**	−4.16	19.63**	−92.46**
Non Occupational	−92.68**	−100.14**	7.92	−0.14
AFQT Percentile	1.93**	1.73**	1.89**	2.70**
AFQT Unknown	93.03**	145.87**	−17.53	193.01**
1980 Cohort[d]	61.26**	−2.74	−198.34**	−59.86**
1982 Cohort	159.27**	−7.90	−236.65**	−53.25**
1984 Cohort	234.25**	−24.20*	−276.85**	−97.37**
1986 Cohort	283.26**	24.45*	−587.90**	−134.06**
1988 Cohort	305.98**	18.85	−691.51**	−86.64**
1990 Cohort	350.33**	−27.96*	−633.88**	−134.38**
R-Squared	.61	.67	.66	.68
Number of Observations	73,003	63,752	25,798	50,168
Mean Monthly Cash Compensation Including Housing	1,920.2	1,854.5	1,884.4	2,029.2

[a]Omitted category is White Male.
[b]Omitted category is 0 Dependents.
[c]Omitted category is Infantry, Combat Arms.
[d]Omitted category is 1978 Cohort.
**Significant at 0.01 level; *Significant at 0.05 level.

Table 5.24

Coefficients of Regression of Monthly Total Compensation (in dollars) on Individual Characteristics, by Service

Characteristic	Army	Navy	Marines	Air Force
Intercept	1,247.89**	1,671.77**	895.28**	896.72**
Years of Service	746.22**	655.82**	867.95**	700.95**
Years of Service Squared	−28.69**	−27.03**	−37.77**	−26.06**
Age at Entry	−23.65**	−29.27*	−17.64**	−26.00**
White Female[a]	−56.78**	−8.53	−30.37	−0.90
Black Male	−138.65**	−166.47**	−94.04**	−183.47**
Black Female	−108.76**	−176.48**	−91.46**	−119.05**
Other Male	12.60	−35.57**	12.10	−62.82**
Other Female	−0.89	−7.10	−26.14	−60.60*
1 Dependent[b]	332.79**	275.85**	317.56**	295.36**
2 Dependents	426.91**	357.95**	453.61**	371.05**
3 Dependents	401.43**	384.03**	489.57**	366.21**
4+ Dependents	374.96**	350.56**	436.35**	369.44**
Electronic Equipment Repairer[c]	−71.26**	74.10**	287.21**	137.50**
Communications/ Intelligence Specialist	−26.23**	89.58**	145.47**	368.60**
Health Care Specialist	−62.76**	−19.01	0.0**	−63.91**
Other Technical and Allied Specialists	8.51	57.61*	195.54**	16.42
Functional Support and Administration	−94.66**	−4.24	129.71**	−39.57**
Electrical/Mechanical Equipment Repairer	−78.15**	−14.09**	183.32**	13.37
Craftsman	−108.97**	30.82**	61.07	−106.15**
Service and Supply Handler	−82.02**	18.66	62.25**	−109.21**
Non Occupational	−99.17**	122.72**	341.66**	79.18
AFQT Percentile	1.34**	2.58**	4.09**	2.82**
AFQT Unknown	317.25**	−57.94**	404.96**	486.34**
1980 Cohort[d]	−392.10**	−389.54**	−207.47**	−296.34**
1982 Cohort	−487.54**	−389.43**	−403.78**	−188.39**
1984 Cohort	−220.18**	−291.98**	−497.14**	−127.65**
1986 Cohort	−180.92**	−254.17**	−622.22**	134.12**
1988 Cohort	83.94**	13.95	−579.96**	731.27**
1990 Cohort	−205.71**	−634.59**	−1,090.03**	67.18
R-Squared	0.84	0.80	0.76	0.70
Number of Observations	73,003	63,752	25,798	50,168
Mean Monthly Total Compensation	4,063.35	3,823.76	3,696.25	4,354.38

[a]Omitted category is White Male.
[b]Omitted category is 0 Dependents.
[c]Omitted category is Infantry, Combat Arms.
[d]Omitted category is 1978 Cohort.
**Significant at 0.01 level; *Significant at 0.05 level.

6. Conclusions

This report has examined how military enlisted personnel are paid in terms of the components that make up compensation and the variation of enlisted compensation with service-member characteristics. First, we found a large degree of variation in both the incidence of receipt and the amount received for different compensation components. While all service members receive Basic Pay, Medical Benefits, the Tax Advantage, and our Retirement Benefit (see Appendix D), four-fifths or less receive each of the following: BAQ, BAS, In-Kind Housing, and the Enlistment/Reenlistment Bonus.

Basic Pay made the largest contribution to Cash Compensation both including and excluding the value of Housing: Basic Pay accounted for 78 percent of the level of Cash Compensation Without Housing and 66 percent of the level of Cash Compensation Including Housing, across all years of service. The other components generally contributed less than 10 percent of the level of Cash Compensation.

In explaining the variance in Cash Compensation measures, Enlistment/ Reenlistment Bonuses and Housing made the largest contributions within years of service of eight or more. Across years of service, each of three components, Basic Pay, Enlistment/Reenlistment Bonuses, and Housing, made up about one-fifth of the variances in Cash Compensation Including Housing. We can infer from this finding that, while Basic Pay determines the preponderance of the levels of Cash Compensation, it does not explain the bulk of differences in members' pay. For Total Compensation, Basic Pay and the Retirement Benefit together accounted for nearly 70 percent of the levels; the Retirement Benefit, Enlistment/Reenlistment Bonuses, and Housing made up nearly all of the variance within a year; and the Retirement Benefit accounted for nearly half the variance across years. Again, this finding implies that service members who are being rewarded differentially are getting the additional compensation through deferred compensation, benefits, and bonuses.

The regression results confirm the finding that almost everyone receives the same Basic Pay. First, the intercept accounted for the largest fraction of Basic Pay, with service-member characteristics explaining little of the rest. Of the characteristics, only years of service made a nonnegligible contribution. More characteristics made a difference in the Cash Compensation measures, with years of service and the presence of dependents making the largest differences. While other characteristics such as race, gender, occupation, and AFQT score made statistically significant differences in compensation measures, their size was generally small, representing less than 5 percent of the mean level of the compensation measures.

Appendix

A. Cohorts Observed for Each Year of Service

Table A.1

Number of Individuals from Each Cohort, Observed at Each Year of Service

YOS	1978	1980	1982	1984	1986	1988	1990
1	3	9	30	70	27,220	26,106	22,045
2	10	41	62	28,069	35,545	35,236	30,345
3	21	82	197	35,500	32,209	31,091	26,917
4	102	290	18,929	28,258	26,939	27,957	23,989
5	286	962	15,729	15,868	14,874	13,527	1,686
6	521	12,636	13,601	13,926	12,775	10,695	0
7	966	14,354	11,908	11,883	9,933	2,069	0
8	9,092	12,915	10,605	10,517	8,486	1	0
9	10,006	10,955	8,934	8,360	1,626	0	0
10	9,013	9,902	8,076	7,053	0	0	0
11	8,088	8,818	6,427	1,321	0	0	0
12	7,591	8,200	5,246	0	0	0	0
13	7,026	6,967	1,039	0	0	0	0
14	6,645	5,669	0	0	0	0	0
15	5,820	1,005	0	0	0	0	0
16	5,132	0	0	0	0	0	0
17	967	0	0	0	0	0	0

NOTE: Analysis file spans 1985–1993.

B. Assignment of Pays to Compensation Components

This table indicates how we allocated the 29 pays reported in the JUMPS file to the component categories we used in the analysis.

Table B.1

JUMPS File–Reported Assignment of Pays to a Compensation Component

Pay Categories Reported in JUMPS File	Compensation Component in Which the Pay Is Included
Basic pay	Basic Pay
Essential-service pay	Special Pay
Foreign-duty pay	Special Pay
Proficiency pay	Special Pay
Overseas-extension pay	Special Pay
Enlistment bonus	Enlistment/Reenlistment Bonus
Selective reenlistment	Enlistment/Reenlistment Bonus
Career sea pay	Special Pay
Career sea pay premium	Special Pay
Hostile-fire pay	Special Pay
Diving-duty pay	Special Pay
Hazardous-duty pays[a]	Special Pay
Basic Allowance for Subsistence (BAS)	BAS
Basic Allowance for Quarters (BAQ)	BAQ
Family Separation Allowance (FSA) 1	Special Pay
FSA 2	Special Pay
Overseas cost-of-living adjustment (COLA)	Special Pay
Overseas Housing Allowance (OHA)	OHA
Variable Housing Allowance (VHA) 1	VHA
VHA 2	VHA
Clothing allowance	Special Pay
Foreign-language pay 1	Special Pay
Foreign-language pay 2	Special Pay

[a]Three separate hazardous-duty pays are reported in JUMPS.

C. Tax-Advantage Calculation

We calculate the tax advantage using the method that is typically used by the Department of Defense. The method takes graduated-tax schedules into account.

First, consider the actuarial method that does not incorporate graduated-tax schedules as the basis for the calculation. Let compensation (COMP) be the sum of taxable benefits (TAXB) and non-taxable benefits (NON): COMP = TAXB + NON. The tax advantage is the before-tax amount that would have to be added to compensation if nontaxable benefits were taxed to make the after-tax compensation equal to the case when the benefits were not taxed. Using this method, the tax advantage (TAXADV) is then

$$\text{TAXADV} = \frac{T \times \text{NON}}{(1-T)} \, , \tag{C.1}$$

where T is the marginal tax rate for an individual's level of TAXB.

We modify this formula to incorporate the fact that tax schedules are graduated. Using the fact that COMP = TAXB + NON, we rewrite the tax-advantage formula as

$$
\begin{aligned}
\text{TAXADV} \ &= \frac{T \times (\text{COMP} - \text{TAXB})}{(1-T)} \\[2mm]
&= \frac{T \times \text{COMP}}{(1-T)} - \frac{T \times \text{TAXB}}{(1-T)} \, .
\end{aligned}
\tag{C.2}
$$

If tax schedules are graduated, it is possible that the tax rate on COMP is not equal to the tax rate on TAXB. Therefore, we compute the tax advantage as

$$\text{TAXADV} = \frac{T_1 \times \text{COMP}}{(1-T_1)} - \frac{T_2 \times \text{TAXB}}{(1-T_2)} \, , \tag{C.3}$$

where T_1 is the tax bracket associated with the level of COMP, and T_2 is the tax bracket associated with the level of TAXB.

D. Retirement-Benefit Calculation

Retirement benefits are an important component of compensation, but computing them is complex. Our approach is to consider the package as an annuity with uncertainty about whether the service member will receive it. Active service members will have different values for these benefits, because the annuity varies in value and because service members may have differing probabilities of reaching the vesting requirement of 20 years. Conditional on receiving a pension, the value of the annuity will depend on several factors: current age, retirement age, expected lifetime, and the benefit stream (monthly payments). To assist in the calculations, we defined some notation for individual i in year t:

B_{it}	=	benefits (annual retirement income in 1992 dollars)
δ	=	discount rate
AGE_{it}	=	age
D_i	=	maximum survival age (assumed to be 84)
R_i	=	retirement age.

Because this annuity is a stream of benefits, we wanted to compute the present discounted value of this stream of benefits (V_{it}) to service member i in year t. To do so, we first assumed the service member completes 20 years of service. In this case, the value of the annuity can be written using the discounted formula:

$$V_{it} = \left[\sum_{s=R_i}^{D_i} \left(\left(\frac{1}{1+\delta} \right)^{s-AGE_{it}} B_{it} \right) \Pr_{it}(\text{live to age } s) \right]. \qquad (D.1)$$

To compute the expected value of this stream of benefits, we needed to also account for the probability (Pr) that the service member will reach retirement with full vesting (20 years of service). Assuming this probability is independent of the expected stream of benefits (EV_{it}),[1] we obtained

---

[1]It may be that those who are more likely to reach retirement may also have higher expected annuities because of faster promotions or other factors that we did not observe. In this case, reaching full vesting and the value of the annuity would not be independent, and we would probably understate the variation in compensation.

$$EV_{it} = V_{it} \bullet \text{Pr}_{it}(\text{reaching } R_i)$$

$$= \left[\sum_{s=R_i}^{D_i} \left(\left(\frac{1}{1+\delta} \right)^{s-\text{AGE}_{it}} B_{it} \right) \text{Pr}_{it}(\text{live to age } s) \right] \bullet \text{Pr}_{it}(\text{reaching } R_i). \quad \text{(D.2)}$$

Members will have different values for *EV* because they differ in their probability of retirement, their expected mortality, and the benefits they can expect after retirement. We now discuss our methods for computing each of these components.

Retirement Benefits, B_{it}

In reality, retirement benefits depend on years of service, retired pay grade, basic-pay history, and disability status. We simplified this computation by assuming that everyone retires after 20 years of service without a disability, and we assumed that they retired with one of three pay grades: E-7, E-8, or E-9. We then identified the basic pay that individuals with 20 years of service would obtain in each of the three grades from basic-pay tables in the *1992 Uniformed Services Almanac* (Ungerleider and Smith, 1992).

Three retirement systems were in effect for our sample of enlisted personnel (see Asch and Warner, 1994, for a summary). Service members who entered before 1980 receive an inflation-protected lifetime annuity equal to (.025*YOS*final basic pay) after completing at least 20 years of service. Service members who entered after 1980 but prior to 1987 receive an annuity of (.025*YOS*[average of highest three years' basic pay]). Those entering in 1987 or later receive a smaller annuity prior to age 62 and the same annuity as 1980–1986 entrants after age 62.

We computed annual retirement benefits according to the formula for members who entered prior to 1980, because we lacked detailed information about the average highest three years' basic pay. Asch and Warner (1994) report that high-three averaging reduces the real value of the annuities by about 6 to 7 percent. Note that high-three averaging will induce a smaller reduction of our estimates of retirement benefits because our estimate is based on the addition to net present values rather than to the retirement-benefit value itself. The true B_{it} that individuals from the 1988 and 1990 cohorts will receive will be even smaller, but this has only minor implications for our estimates, since these cohorts are included for only a few early years of retirement-benefit calculations and since the discounting makes these estimates very small.[2]

---

[2]Another factor we did not include in our calculations is that survivors of a military retiree may continue to collect a portion of the retiree's pension after the recipient dies.

Retirees and their dependents are also eligible for the Civilian Health and Medical Program of the Uniformed Services (CHAMPUS) health insurance benefits, so we also added the value of a family plan into B_{it}. The following table shows these computations.

		Retirement Benefits (B):	
Grade at time t	Retirement Grade	Pension Benefits (1992 $)	Health Benefits (1992 $)
E-7 or less	E-7	24,059	5,292
E-8	E-8	26,960	5,292
E-9	E-9	30,913	5,292

Survival Probabilities, Pr_{it} (live to age s)

Retirement benefits accrue until a member dies. To estimate the probability of survival, we used mortality data (life tables) from the Office of the Actuary, U.S. Department of Defense (1993), for males and *Vital Statistics of the United States, 1990* (National Center for Health Statistics, 1994) for females. Survival probabilities vary by gender only.

Retirement Probabilities, Pr_{it} (reaching R_i)

Much of the variation in *EV* arises because of variation in the probability that enlisted personnel reach 20 years of service. We computed this probability as follows: We first considered a service member i who has just enlisted. The probability that member i reaches retirement can be decomposed into the product of the probabilities that i stays another year in the service; i.e.,

$$
\begin{aligned}
\text{Pr}(i \text{ retires}) &\equiv \text{Pr}(\text{YOS}_i > 0 | i \text{ enlisted}) \cdot \text{Pr}(\text{YOS}_i > 1 | \text{YOS}_i \geq 1) \\
&\quad \cdot \ldots \cdot \text{Pr}(\text{YOS}_i > 20 | \text{YOS}_i \geq 19) \\
&\equiv (1 - \lambda_1) \cdot (1 - \lambda_2) \cdot \ldots \cdot (1 - \lambda_{20}).
\end{aligned}
\tag{D.3}
$$

Each λ_j, on the right-hand side, is the *hazard* that individual i will leave the service after j years, given that i has already served $(j - 1)$ years. For a member i in year t who has already served YOS_{it} years in the military, we can generalize and concatenate the above formula to reflect years already served:

$$
\text{Pr}(R_{it} = 1) \equiv \prod_{s=\text{YOS}_{it}}^{20} (1 - \lambda_s).
\tag{D.4}
$$

We allowed each hazard to depend on characteristics of the service member. Thus, λ_s is assumed to be a function of race (Hispanic, black, white, other), age at enlistment, gender, service (Army, Air Force, Marines, Other), and AFQT score. We estimated $\lambda_1,...,\lambda_{20}$ separately, using maximum likelihood under the assumption of normality (probit model). Thus, we obtained

$$\lambda_{is} \equiv \Phi(x_i\beta_s), \text{ for service member } i \text{ and } s = 1,...,20 \qquad \text{(D.5)}$$

where Φ is the cumulative distribution function for a normally distributed random variable. To compute λ_{is}, we made a subset of our data the sample of personnel who served s years or more. We then defined a binary variable β_s that indicates whether the service member served *exactly* s years. We then computed maximum-likelihood estimates for β_s by estimating a probit model regression of this binary variable on the characteristics noted earlier. We did this computation separately for each of the 20 hazard functions.[3]

This strategy allowed us to identify differences in the likelihood of retention as a function of sociodemographic status and service. Each individual i has a separate set of predicted hazards. To compute the probability of retirement for any individual i in year t, we merely inserted our fitted probabilities into Equation (D.4):

$$\hat{\Pr}(R_{it} = 1) \equiv \prod_{s=\text{YOS}_{it}}^{20} \left(1 - \hat{\lambda}_{is}\right). \qquad \text{(D.6)}$$

Using this equation, we were able to simulate the probability that any individual i reaches retirement after s years of service. Figure D.1 shows the variability in these probabilities in our data. The solid line in Figure D.1 can be considered the "average" retirement path. Because we are interested in variations in compensation as well as the mean, we also show the dispersion in these fitted probabilities by plotting the minimum and maximum fitted retirement probabilities at each year of service. Thus, the segment between the dashed lines at any year of service shows the range of fitted retirement probabilities.

---

[3]We used the 1985 Survey of Officers and Enlisted Personnel (SOEP) to estimate the probabilities of retirement for years 16 through 20, using the same specification as that used for years 1 through 15. The 1985 SOEP is a comprehensive survey of military personnel that addresses issues of importance to service members. The 1985 data also include demographic and career progression data on each respondent from 1985 through 1991. It is the set of data on individuals with 16 to 20 years of service from 1985 through 1991 that we used to estimate the probabilities. The sets of probit estimates, 20 in all, are available from the authors upon request.

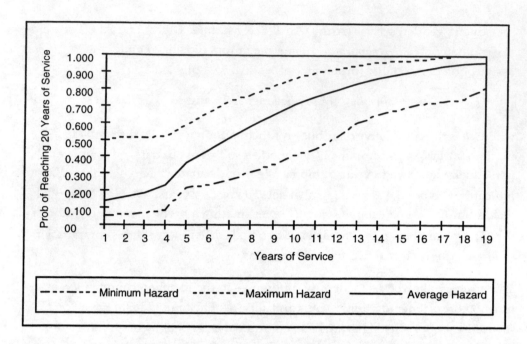

Figure D.1—Dispersion in the Probability of Retirement

E. Variable Definitions

Table E.1

Definition of Variables

Variable	Definition
Years of Service	Number of years the individual has served on active duty
Age at Entry	Age at which the individual enlisted
White Female	White female; may include Hispanics
Black Male	Black male; may include Hispanics
Black Female	Black female; may include Hispanics
Other Male	Nonwhite, nonblack male; may include Hispanics
Other Female	Nonwhite, nonblack female; may include Hispanics
1 Dependent	Service member has one dependent
2 Dependents	Service member has two dependents
3 Dependents	Service member has three dependents
4+ Dependents	Service member has four or more dependents
Electronic Equipment Repairer	Individual's primary Military Occupational Specialty (MOS) is in DoD one-digit occupational group 1[a]
Communications/Intelligence Specialist	Individual's primary MOS is in DoD one-digit occupational group 2
Health Care Specialist	Individual's primary MOS is in DoD one-digit occupational group 3
Other Technical and Allied Specialist	Individual's primary MOS is in DoD one-digit occupational group 4
Functional Support and Administration	Individual's primary MOS is in DoD one-digit occupational group 5
Electrical/Mechanical Equipment Repairer	Individual's primary MOS is in DoD one-digit occupational group 6
Craftsman	Individual's primary MOS is in DoD one-digit occupational group 7
Service & Supply Handler	Individual's primary MOS is in DoD one-digit occupational group 8
Non Occupational	Individual's primary MOS is in DoD one-digit occupational group 9
AFQT Percentile	Person's percentile score on AFQT
AFQT Unknown	Person's percentile score on AFQT is not known
1980 Cohort	Individual enlisted in 1980
1982 Cohort	Individual enlisted in 1982
1984 Cohort	Individual enlisted in 1984
1986 Cohort	Individual enlisted in 1986
1988 Cohort	Individual enlisted in 1988
1990 Cohort	Individual enlisted in 1990

[a]For more information on these codes, see U.S. Department of Defense, *Occupational Conversion Index,* Washington, D.C., 1993.

F. Regression Results for BAS

Table F.1

Coefficients of Regression of Monthly BAS (in dollars) on Individual Characteristics, by Service

Characteristic	Army	Navy	Marines	Air Force
Intercept	−27.16**	10.84**	−38.46**	58.60**
Years of Service	22.16**	18.84**	15.18**	19.65**
Years of Service Squared	−0.98**	−0.98**	−0.58**	−1.01**
Age at Entry	−0.11	−0.09	0.92**	1.13**
White Female	26.11**	53.26**	32.13**	15.12**
Black Male	−6.77**	−5.30**	−7.96**	−4.70**
Black Female	15.48**	46.69**	22.41**	6.27**
Other Male	−3.04**	−2.26**	−4.08**	−1.78**
Other Female	22.43**	47.97**	28.91**	8.30**
1 Dependent	65.75**	15.08**	77.43**	45.68**
2 Dependents	73.59**	21.46**	87.83**	44.31**
3 Dependents	75.13**	20.21**	87.08**	40.22**
4+ Dependents	74.61**	22.12**	83.45**	36.66**
Electronic Equipment Repairer	9.21**	3.46**	12.16**	−9.20**
Communications/ Intelligence Specialist	6.17**	8.70**	9.55**	−1.73
Health Care Specialist	16.27**	46.26**	—	−11.49**
Other Technical and Allied Specialist	6.02**	24.23**	29.69**	0.57
Functional Support and Administration	12.00**	18.64**	19.32**	−10.60**
Electrical/Mechanical Equipment Repairer	4.82**	7.40**	10.45**	−13.44**
Craftsman	4.10**	−6.12**	12.40**	−19.06**
Service and Supply Handler	14.11**	−4.25**	11.27**	−12.14**
Non Occupational	−8.33**	−11.73**	5.17**	−73.16**
AFQT Percentile	0.16**	0.22**	0.17**	0.04**
AFQT Unknown	6.37**	8.84**	2.52	2.00
1980 Cohort	−4.87**	0.69	−0.06	−5.92**
1982 Cohort	0.66	−5.27**	−2.96*	−5.43**
1984 Cohort	4.12**	−7.40**	−6.93**	−6.36**
1986 Cohort	0.88	−11.85**	−10.19**	−10.35**
1988 Cohort	1.94*	−17.67**	−12.46**	−14.41**
1990 Cohort	9.96**	−23.36**	−8.14**	−22.40**
R-Squared	0.35	0.25	0.42	0.27
Number of Observations	313,022	223,651	98,525	211,771
Mean Monthly BAS	101.41	96.97	80.18	156.36

[a]Omitted category is White Male.

[b]Omitted category is 0 Dependents.

[c]Omitted category is Infantry, Combat Arms.

[d]Omitted category is 1978 Cohort.

References

Amemiya, Takeshi, *Advanced Econometrics*, Cambridge, Mass.: Harvard University Press, 1985.

Angrist, Joshua D., "Lifetime Earnings and the Vietnam Era Draft Lottery: Evidence from Social Security Administrative Records," *The American Economic Review*, Vol. 80, No. 3, 1990.

————, "Using the Draft Lottery to Measure the Effect of Military Service on Civilian Labor Market Outcomes," *Research in Labor Economics*, Vol. 10, 1989 (ed. Ronald Ehrenberg, JAI Press).

Asch, Beth J., *Designing Military Pay: Contributions and Implications of the Economics Literature*, Santa Monica, Calif.: RAND, MR-161-FMP, 1993.

Asch, Beth J., and James N. Dertouzos, *Educational Benefits Versus Enlistment Bonuses: A Comparison of Recruiting Options*, Santa Monica, Calif.: RAND, MR-302-OSD, 1994.

Asch, Beth J., and John T. Warner, *A Policy Analysis of Alternative Military Retirement Systems*, Santa Monica, Calif.: RAND, MR-465-OSD, 1994.

Becker, Gary S., *Human Capital: A Theoretical and Empirical Analysis, with Special Reference to Education*, 2nd Ed., New York: Columbia University Press for the National Bureau of Economic Research, 1975.

Berger, Mark C., and Barry T. Hirsch, "The Civilian Earnings Experience of Vietnam-Era Veterans," *Journal of Human Resources*, Vol. 8, No. 4, 1983, pp. 455–479.

Black, Matthew, Robert Moffitt, and John T. Warner, "The Dynamics of Job Separation: The Case of Federal Employees," *Journal of Applied Econometrics*, Vol. 5, 1990, pp. 245–262.

Bureau of Labor Statistics, *Employee Benefits in Medium and Large Private Establishments, 1993*, Washington, D.C.: U.S. Government Printing Office, Bulletin 2456, 1994.

Camm, Frank A., *Housing Demand and Department of Defense Policy on Housing Allowances*, Santa Monica, Calif.: RAND, R-3865-FMP, September 1990.

Daniel, Kermit, "Does Marriage Make Workers More Productive?" Philadelphia, Penn.: The Wharton School, University of Pennsylvania, Working Paper, 1994.

Gotz, Glenn A., and John J. McCall, "Sequential Analysis of the Stay/Leave Decision: U.S. Air Force Officers," *Management Science*, Vol. 29, No. 3, March 1983, pp. 335–351.

74

Greene, William H., *Econometric Analysis*, 2nd Ed., New York: Macmillan Publishing Company, 1993.

Hogan, Paul F., and Matthew Black, "Reenlistment Models: A Methodological Review," in Curtis L. Gilroy, David K. Horne, and D. Alton Smith, eds., *Military Compensation and Personnel Retention: Models and Evidence*, Alexandria, Va.: U.S. Army Research Institute for the Behavioral and Social Sciences, February 1991.

Hosek, James, and Christine Peterson, "Enlistment Decisions of Young Men," in Curtis Gilroy, ed., *Army Manpower Economics*, Boulder, Colo., and London: Westview Press, 1985.

—————, *Serving Her Country: An Analysis of Women's Enlistment*, Santa Monica, Calif.: RAND, R-3853-FMP, 1990.

Hosek, Susan D., et al., *The Demand for Military Health Care: Supporting Research for a Comprehensive Study of the Military Health Care System*, Santa Monica, Calif.: RAND, MR-407-1-OSD, 1995.

Lazear, Edward P., "Why Is There Mandatory Retirement?" *Journal of Political Economy*, Vol. 87, No. 6, 1979, pp. 1261–1284.

Lillard, Lee, "Inequality: Earnings vs. Human Wealth," *American Economic Review*, Vol. 67, No. 2, 1977, p. 42–53.

Mincer, Jacob, *Schooling, Experience, and Earnings*, New York: National Bureau of Economic Research, 1974.

Moffitt, Robert, "Estimating the Value of an In-Kind Transfer: The Case of Food Stamps," *Econometrica*, Vol. 57, No. 2, March 1989, pp. 385–409.

Murphy, Kevin, and Finis Welch, "Empirical Age-Earnings Profiles," *Journal of Labor Economics*, Vol. 8, No. 2, 1990, pp. 202–229.

Murray, Michael P., "Real Versus Monetary Transfers: Lessons from the American Experience," in M. Pfaff, ed., *Public Transfers* and *Some Private Alternatives During the Recession*, Berlin: Duncker and Hunblot, 1983, pp. 105–123.

National Center for Health Statistics, *Vital Statistics of the United States, 1990*, Washington, D.C.: U.S. Public Health Service, 1994.

O'Neill, June, "The Role of Human Capital in Earnings Differences Between Black and White Men," *Journal of Economic Perspectives*, Vol. 4, No. 4, 1990, pp. 25–45.

Phillips, Douglas W., and David A. Wise, "Military Versus Civilian Pay: A Descriptive Discussion," in David A. Wise, ed., *Public Sector Payrolls*, Chicago: The University of Chicago Press, Project Report, National Bureau of Economic Research, 1987.

Rosen, Sherwin, "The Military as an Internal Labor Market: Some Allocation, Productivity, and Incentive Problems," *Social Science Quarterly*, Vol. 73, No. 2, June 1992, pp. 227–237.

———, "The Theory of Equalizing Differences," in O. Ashenfelter and R. Layard, eds., *Handbook of Labor Economics*, Vol. 1, New York: North-Holland, distributors for Elsevier Science Publishers, 1986.

Rosen, Sherwin, and Paul Taubman, "Changes in Life Cycle Earnings: What Do Social Security Data Show?" *Journal of Human Resources*, Vol. 17, Summer 1982, pp. 321–338.

Sloss, Elizabeth M., and Susan D. Hosek, *Evaluation of the CHAMPUS Reform Initiative*, Volume 2, *Beneficiary Access and Satisfaction*, Santa Monica, Calif.: RAND, R-4244/2-HA, 1993.

Smeeding, T., "The Antipoverty Effectiveness of In-Kind Transfers," *Journal of Human Resources*, Summer 1977, pp. 360–378.

Smith, MSG Gary L. (Ret), and Debra M. Gordon, *1994 Uniformed Services Almanac*, Falls Church, Va.: Uniformed Services Almanac, Inc., 1994.

Sullivan, C., M. Miller, R. Feldman, and B. Dowd, "Employer-Sponsored Health Insurance in 1991," *Health Affairs*, Vol. 11, No. 4, 1992, pp. 172–185.

Topel, Robert, "Specific Capital, Mobility, and Wages: Wages Rise with Job Seniority," *Journal of Political Economy*, Vol. 99, No. 1, 1991.

Ungerleider, Col. Al (Ret), and MSG Gary L. Smith (Ret), *1992 Uniformed Services Almanac*, Falls Church, Va.: Uniformed Services Almanac, Inc., 1992.

U.S. Bureau of the Census, *Statistical Abstract of the United States: 1993*, 113th Ed., Washington, D.C., 1993.

U.S. Department of Defense, Office of the Secretary of Defense (OSD), *Military Compensation Background Papers: Compensation Elements and Related Manpower Cost Items: Their Purposes and Legislative Backgrounds*, 4th Ed., Washington, D.C.: U.S. Government Printing Office, November 1991.

U.S. Department of Defense, Office of the Actuary, Office of the Assistant Secretary of Defense, *Valuation of the Military Retirement System*, Washington, D.C., September 30, 1993.

U.S. Department of Defense, Office of the Assistant Secretary of Defense, Personnel and Readiness, *Occupational Conversion Index, Enlisted/Officer/Civilian*, Washington, D.C., DoD 1312.1-I, September 1993.

Willis, Robert, and Sherwin Rosen. "Education and Self-Selection," *Journal of Political Economy* Vol. 87, No. 5, Part 2, 1979, pp. S7–S36.